Variability of Climate Change and Forest

Variability of Climate Change and Forest

Anderson, Max Follis

RANDOM

RANDOM PUBLICATIONS
NEW DELHI (INDIA)

Variability of Climate Change and Forest

978-93-5111-169-6

Published in 2013 in India by

RANDOM PUBLICATIONS

4376-A/4B, Gali Murari Lal, Ansari Road
New Delhi-110 002
Phone : +9111-43580356, 011-23289044
e-mail: randompublications@hotmail.com

Reprinted 2026

Type Setting by : Friends Media, Delhi-110089

Contents

Preface

Climate change is emerging as one of the most important issues of our time, with the potential to cause profound cascading effects on ecosystems and society. However, these effects are poorly understood and our projections for climate change trends and effects have thus far proven to be inaccurate.

Climate change is a significant and lasting change in the statistical distribution of weather patterns over periods ranging from decades to millions of years. It may be a change in average weather conditions or the distribution of events around that average (e.g., more or fewer extreme weather events). Climate change may be limited to a specific region or may occur across the whole Earth. Another significant climate trend that has been linked to systematic changes in temperature and precipitation is an increase in the variability, or extremes, of climate. This change in extremes can alter not only the intensity of individual events, such as storms and floods, but the timing and spatial patterns of weather as well. Since the mid-1800s the average surface temperature of the globe has risen about 0.4°C-0.6°C, and periods of persistent warming can, in general, be associated with increased variability.

One of the more challenging problems for eco-physiologists is to assess the supply of nutrients available to sustain forest growth. A little insight is gained by simply analysing the elemental content in the soil because many nutrients are present in unavailable forms, and trees differ in their abilities to extract essential nutrients from throughout the soil profile. In addition, a range of nutrients can be acquired from the air via aerosol deposition on foliage and additional nitrogen can be acquired through root associations with symbiotic nitrogen-fixing bacteria. Physiologists and chemists, starting with Liebig in 1840, have established those elements essential for growth, defined specific deficiency symptoms and related concentrations in foliage to minimum requirements. We emphasize two additional components of nutrition that have significant impact on the growth of forests: balance and internal recycling.

This scholarly work is both broad in scope and extremely rich in detail. The readers will find this a useful reference book as well as a general text.

—*Author*

1

Variability of Climate Change

VARIABILITY

Variability — that is the outstanding characteristic of temperate-zone weather. Even the brief sea log just quoted shows how weather in the tropics is much the same from place to place and day to day; temperate-zone weather, conversely, is diverse and changeable. But had our voyage extended far enough northward, it would have reached another region of comparative weather sameness — cold sameness instead of warm sameness -the arctic zone.

The temperate zone can have no stable and typical weather of its own. To southward (in the northern hemisphere) is the steady, warm, moist climate of the tropics. To northward is the fairly steady, mostly cold, fairly dry climate of the arctic. (This picture is of course repeated, with reverse directions, in the southern hemisphere.) The temperate zone is a buffer region between tropic and arctic climates, a battleground on which the gigantic atmospheric forces of heat and moisture advance and retreat, an unstable compromise between irreconcilable extremes, swayed this way and that in its evanescent weather as tropic forces (warm air masses) or arctic forces (cold air masses) momentarily gain ascendancy.

TROPIC CLIMATE

Tropic wind-and-climate belts migrate somewhat, north and south, with the apparent north-south seasonal swing of the sun. To get within-the-tropics weather in its most typical form we must go well south of the Tropic of Cancer, nearly down to the equator — say to Panama in about lat. 8° N. In these latitudes there are only two main seasons — a winter dry season under the influence of the steady northeast trades, and a summer wet season brooded over by the fitful calms of the doldrums.

The outstanding features of tropical-marine climate are that the average temperature is practically constant from month to month (nor does it ever vary, by more than a few degrees, from day to day); and that the cloudiness and rainfall, always appreciable, reach high values during the wet-season doldrums invasion of summer and early fall.

Upper winds over Panama, at half-mile altitude, average northeast about 8 m.p.h. in the rainy season and north-northeast about 20 m.p.h. in the dry. At two miles altitude there is less seasonal change — the average wind only varies from east 10 m.p.h. (rainy) to cast-northeast 12 m.p.h. (dry). Upper-air characteristics over the Caribbean Sea, quite typical of tropical-marine climate in general, were studied during the 1939 hurricane season by a United States Weather Bureau expedition to Swan Island.

This four-mile-long islet, named after a nineteenth-century pirate, lies almost exactly in the middle of the Sea at about lat. 16° N., between Jamaica and British Honduras. Within four months, observers Rhamlow and Leech released some two hundred and fifty pilot balloons and one hundred and twenty radio-sounding balloons. They found that the trade wind extended well aloft, though with diminishing force, here as in Panama — prevailing easterlies all the way up to about two and a half miles altitude. But above this level, the upper winds are mostly out of the west and increase with altitude.

Surface temperatures at Swan Island, and all over the Caribbean area, average about +75° to 80° F. the year round. But at the base of the stratosphere, ten miles up in these latitudes, it is cold indeed — about — 105° to — 110° F. Thus in the thick tropical troposphere there is an average lapse rate of something like — 3½° F/1000 ft (about — 6° C/Km). Together with plenty of warmth and moisture in the lower levels, this lapse rate means that conditional instability must extend through considerable air layers, particularly in the vertical-sun-warmed rainy season of summer and fall — and it is this conditional instability, together with inexhaustible supplies of moisture from the warm ocean, that produces the daily torrential showers and thundershowers, the regular afternoon downpours and the moist, misty mornings which can make equatorial regions such a continual sweat-bath for ill-adapted white men.

On or close to the ocean this deep-tropical weather is bearable, or perhaps even comfortable. But the steaming jungles inland, where temperatures can rise unchecked and where all breeze is choked off by the tangled uprearing of dense vegetation, are sometimes aptly described as 'green hells.' As Julian Duguid wrote '... a rich eternal garment of green, dappled with golden sunspots... a dense, fever-stricken thicket shimmering in the heat with a perpetual glassy haze....' However, by no means all the land areas within the tropics are covered with jungle — some of them are sandy and arid in the extreme, and doubly hot though not so muggy. (The word 'jungle' itself, originally, meant only 'waste land' in India.) And even in the fecund jungle areas of the tropics, such as Panama and northern South America, vegetation waxes lush and green with the rainy season, wanes sparse and tawny with the dry season.

There are also irregular year-to-year variations in moisture and prevailing winds that change the face of tropical nature. I got off a ship on the Colombian coast in August, 1939, and walked inland for a while, expecting to wade through wetseason green tangles of vines and trees peopled with tropic creatures — but the soil proved to be dry and sandy, the jungle a dried-up shrinkage of scrub, the creatures hidden away from the glaring vertical sun.

ARCTIC CLIMATE

Before Admiral Peary of the United States Navy marched five hundred icy miles on snowshoes from Ellesmere Land to the North Pole in the early spring of 1909, and since his time, many expeditions have entered the northern arctic regions. No latter-day explorer, incidentally, has ever surpassed Peary's record of intelligent planning and courageous, competent performance. But the airplane has rung startling changes in arctic travel.

In May, 1937, four large air transports set down four Russian scientists and a dog, together with a little portable cabin and food and fuel for a year or more, on one of the uncertain drifting ice floes that cover the Arctic Ocean around the pole. Leader Papanin, astronomer-aerologist Federov, hydrologist Shirsov, and radioman Krenkel had hoped to remain at or near the earth's axis for a whole year of observations, but their ten-foot-thick ice floe soon began to drift (as Peary or Fridtjof Nansen could have foretold), at first very slowly, across the two-mile-deep polar basin towards Greenland.

Then it moved, less gradually, southward along the east coast of that high continent, and finally sailed at twenty-five miles a day on southward. The floe-campers were rescued by icebreakers and airplanes twelve hundred miles south of the pole in February, 1938. Papanin's expedition found the summer temperatures of the polar basin hovering mostly around + 30° to 46° F.; clear spells of a day or two alternated with dull, overcast, rainy-to-wetsnowy weather. Winds varied from calm to light or moderate, mostly out of the north or northeast. Much of the ice was melting in the warmer spells, adding to the troubles and perils of the ice-campers. There was always the danger that their floe would disintegrate and dump them unceremoniously into the Arctic Ocean.

When winter came they had drifted several hundred miles south of the pole itself, but winter climate around the polar basin is probably pretty much the same everywhere. Temperatures ranged all the way from 50° F. below zero to 30° or more above, in accordance with the sweep of unlike air masses and cyclones, which occasionally extends well into the Arctic. For several months in midwinter the polar region is, of course, in continual darkness, with the sun visible at first in the spring only as a pale horizon glow around noon. On the thirty-first of December, strangely enough, the Russian wind-powered radio reported a temperature of + 43° F. over the arctic ice! But within a few hours, the mercury had tobogganed to a more usual value of -31° F.

The general north-polar region is, of course, about midway on the most direct (great-circle) air routes between Europe and Asia, and between eastern United States and Asia. Some years hence, perhaps, great 500-mile-an-hour strato-transports will be winging daily across these icy-watery wastes. Compared with other climatic regions, the polar area offers good flying conditions — during at least two or three days out of a week, that is, and particularly at higher levels.

Spring, when the polar high-pressure area lies symmetrically over the geographic pole, and not over the continental 'cold poles' in Greenland and northern Siberia, is the best flying season. A sealed-cabin, supercharged stratoliner, in any case, would be well up in the low arctic stratosphere (above thirty thousand feet) most of the time. Aside from Ellesmere Land at the north tip of America, one of the best fixed bases for observing the climate of the north-polar basin is the small island of Franz Joseph Land in lat. 81° N. long. 60° E., about six hundred land miles from the pole. In 1932-33 the Russian scientist Moltchanov (incidentally one of the pioneers in the development of radio-sounding balloons) made both surface and upper-air observations on Franz Joseph Land for a year. Like the floe-campers, he found average summer surface temperatures around +30° to 40° F. Winter temperatures averaged around -10 F.

Above the ground, for the first mile, there was usually a temperature inversion amounting to 5° F. or so; above that height the lapse rate or vertical temperature gradient was always surprisingly uniform at -3° or -4° F. per thousand feet up to the low polar 'tropopause,' marking the beginning of the stratosphere at about five miles altitude in winter, six miles in summer. Temperatures at the stratosphere base averaged about -75° F. in winter, about -40° F. in summer. July and August were the warmest months, as usual in the northern hemisphere; December was the coldest month at the surface, February coldest aloft. The southern hemisphere in general, being mostly water, shows more typical wind-and-climate belts than the northern hemisphere, which is all cluttered up with land masses that get too warm in summer and too cold in winter. By the same token, or a reverse one, the antarctic climate, founded upon a high, icecovered continent symmetric around the geographic pole, is more typically 'polar' than the arctic climate.

During the 1933-35 Byrd occupancy of 'Little America' in about lat. 79° S., long. 164° W., aerologists Haines and Grimminger recorded a great deal of antarctic weather. Mostly the average temperature was around +20° to 25° F. in summer (January), and around -35° to -40° in winter (July) — considerably colder, that is, than the north-polar climate. The coldest summer temperature was +12° on January 1; the warmest, +37° on December 22. In winter the mildest reading was +23° on July 27; the coldest was -71° on July 21. During a roaring blizzard in July the temperature fell to -65° with the wind sixty miles an hour out of the southeast — and wind, rather than temperature, is what really hurts in cold climates. These southeasterly blizzards in Antarctica correspond to the northeasterly gales around the north-polar basin.

High winds in Antarctica sometimes come from the north, just as north-polar gales can arise out of the south, but the colder blizzards from somewhere around the South Pole are more usual. Laurence M. Gould, Byrd's geologist and deputy in 1928-30, writes of one terrific southerly gale in the Rockefeller Mountains that first blew away the heavily packed snow wall in front of his firmly anchored airplane, and then finally blew the plane itself away and wrecked it. Some time before the maximum wind came, the airplane's air-speed metre indicated a steady velocity of ninety miles per hour, with gusts well over a hundred.

CONQUEST OF CLIMATE

The decade from 1900 to 1910 signalized an. amazing conquest of the tropics as a livable environment for civilized man. The years after 1940, likewise, may see a parallel conquest of the arctics. Within the tropics, except in certain areas plagued by extreme and wearisome heat, general living conditions were easy enough — too easy for man's own good, in fact; and aside from air transport, radio communication, mechanical refrigeration, and other new conveniences, the conquest was nearly all medical. Such dread warm-country diseases as typhoid, cholera, and yellow fever were finally vanquished — at least in the case of selected, well policed areas.

The work of Walter Reed, his assistant surgeons, and the heroic American soldiers who voluntarily entered the valley of 'yellow jack's' macabre shadow -all this really conquered the tropics and equatorial regions for white men, and contributed more than engineering and steam shovels to the triumphant completion, in 1913, of the world's greatest inter-ocean canal. Within the arctics and around the poles, on the contrary, disease is unknown unless brought to the scene by human carriers, but living conditions (without many bothersome precautions) are unduly hard and exacting. Aside from the extremes of dark and cold and gale and blinding snow, the chief hazards are: difficulty in finding your way around (particularly in the long, cold, dark arctic night); difficulty in surviving ordinary mischances such as falling into water, badly tearing clothes, and so on, particularly if alone; difficulty in getting adequately balanced diet and sufficient ultra-violet radiation.

Yet even now, with our immense progress in transportation and communication, some of these difficulties can be largely overcome at a properly organized arctic station. Clothing must, of course, be of arctic type — layers of wool or fur inside for warmth, layers of canvas or leather outside for wind-breaking. Adequate shelter, plenty of fuel, and fresh food can now be transported easily by airplane to almost any arctic site. On location, howling arctic gales will provide plenty of electric power for lighting and radio communication, and perhaps for heating as well.

Many of the dark terrors of the outer arctic night, which have heretofore held men prisoned for months in their base camps (except in the full of the moon) can perhaps be conquered by such devices as more powerful and efficient portable searchlights (using super-hot filaments, say, or possibly fluorescent light-sources); portable radio-telephone sets for constant two-way communication with and direction-finding on one or more base stations; more compact and accurate instruments, and simpler methods, for celestial navigation.

Against the white, too-uniform snow-glare of the polar day, improved goggles (perhaps of polarizing type) may be employed. Lighter, stronger and warmer tents and clothing can perhaps be devised. Much of the heavy local leg-work of arctic exploration can now be done with snow-mobiles, either tread-driven or propeller-driven. By means of vitamin diets, ultra-violet sunlamps and radio-diathermy, human beings near the poles will perhaps be enabled to maintain a high state of health during the long arctic night, and to overcome quickly any respiratory diseases that do develop.

Eventually we may even see arctic health resorts, rest camps and play camps. Ellsmere Land in northern Canada (Peary's jumping-off place), less than three hours' flying time from the North Pole, is only twelve hours by present slow air liner from New York City. 'Polar Dude Camp on Ellsmere Land — spend next weekend here in the dark restfulness of the polar night under the flickering curtain of the glorious aurora borealis; pack ice exploring parties for the more active' — and so on. Such an advertisement may not startle our children twenty years hence.

TEMPERATE CLIMATE

We have said that the temperate zone is, in a climatic sense, nothing better than a battleground between tropic and arctic weather forces — between warm and cold air masses. Nevertheless the temperate zone (and in particular, the north-temperate zone) is that part of the earth in which civilized man evolved, and is probably the region most favorable to his continued progress and development. On the *average,* of course, temperate temperatures are pleasantest and best for all human activities — neither too hot nor too cold — though this average may rarely be matched by the vagaries of daily weather. And even the extreme moodiness and changeability of temperate weather has its advantages. Variety in weather, at least as to the four- or five-day spells and sequences so common in middle latitudes, perhaps satisfies some deep, innate rhythmic urges in man which cannot be allowed to atrophy without intellectual and cultural and physical loss.

However such obscure causes may or may not operate, it is certain that most human progress up from the dumb-ape level — most human art and ingenuity and imagination, most human science and engineering and social organization — has been best nurtured so far, and probably will be best nurtured in the future, in temperate latitudes. Among all the various meteorological factors that go to make up a climate, temperature is the one that most directly influences human health, happiness, and efficiency. How do various temperatures really feel to average human beings? Different degrees of heat of course feel quite different to different people, or to the same people under different conditions — in particular, under varying conditions of exertion. But it is possible to indicate average human reactions to temperature, and they are somewhat as follows: High humidity makes warm temperatures feel warmer, by retarding evaporation from the body. High humidity also makes cold temperatures feel colder, by making the clothing conduct heat better away from the body.

Wind makes cold temperatures feel colder, by removing the cushion of warm air that tends to form around the body in the absence of wind. This wind-cooling effect is particularly powerful at below-zero temperatures. On the other hand, wind makes warm temperatures up to 100° F. feel less warm, by increasing evaporation from the body. But if the temperature is over 105°, wind increases the discomfort.

The optimum temperature range, 68°-70° F., feels more comfortable if the air is normally moist. For health and comfort, indoor relative humidity should be around sixty per cent. But in the average home or office, in winter, it is seldom more than twenty per cent. Evaporating pans on radiators help some, but not very much. Constant-duty electric evaporators, preferably combined with electric lights, might well be used in temperate-zone houses during the winter.

Today's temperature anywhere is indicated by the thermometer; tomorrow's (which in temperate or arctic latitudes may be shockingly different) is foretold by a good forecast. Beyond these limits, our best present guide to future temperatures is the 'normal.' To get the normal temperature for any particular day of the year at any particular place, the average temperatures for that day, observed over as long a span of years as possible, are all averaged together. The normal temperature is, of course, merely a fictitious quantity seldom or never duplicated by any actual daily average; for climatically speaking, we dwellers in the temperate zone are always traveling now south, now north with the ebb and flow of warm and cold air masses.

Is our North American temperate-zone climate changing? This question keeps recurring in weather talk, and most of the oft-quoted answers foist off something about 'colder winters in Grandfather's day'— a notion perhaps promoted by oldsters in support of their claims to general superiority. The rising generation, contrariwise, have been inclined to charge the whole idea off to dotage and general boastfulness.

A definitive answer has been given by J. B. Kincer of the United States Weather Bureau: For a good many years up through 1936 there has been a decided tendency to warmer and drier weather, the trends being especially marked during the past quarter of a century, notwithstanding an occasional bad flood or a severely cold winter. Take the winter season, for example: With the exception of those for 1917-18 and 1935-36, the winters for the past twenty-five years or more, considering the country as a whole, have been rather uniformly warmer than normal, and, on an annual basis, every year since 1929 has had above normal temperature.

Also, in the matter of rainfall, there has been an equally marked tendency to droughts in recent years. However, an examination of the longer weather records of the country, going back one hundred years or more, indicates that this does not represent a permanent change of climate, but rather a warm, dry phase of our normal climate, to be followed, doubtless, by a cooler, wetter phase, when there will be more rain in summer and lower temperatures in winter....

Since the turn of the century, Mr. Kincer goes on to say, there has been a distinct yearly trend towards warmer winters and drier summers all over the world — and this in spite of a few exceptionally cold winters here and there. In other words, there are long (say fifty-year) cycles in climate, just as there are short (say five-day) cycles in weather. If all these cycles were only considerate enough to maintain constant length, amplitude, and form, both weather forecasting and climate forecasting would be child's play. But the form, intensity, and length of weather and climate changes are all complexly variable. In any case the semipermanent (say 500-year) picture of our American climate shows, apparently, no appreciable change. But the geologic-time picture, measured in thousands or millions of years, would again show marked climatic cycles — the advance and retreat of the various ice ages.

SIGNIFICANT OF CLIMATE CHANGE

Climate change is a significant and lasting change in the statistical distribution of weather patterns over periods ranging from decades to millions of years. It may be a change in average weather conditions or the distribution of events around that average (*e.g.*, more or fewer extreme weather events). Climate change may be limited to a specific region or may occur across the whole Earth.

KINDS OF CLIMATE CHANGE

Agriculture

Approximately 2.4 billion people live in the drainage basin of the Himalayan rivers. India, China, Pakistan, Afghanistan, Bangladesh, Nepal and Myanmar could experience floods followed by severe droughts in coming decades. In India alone, the Ganges provides water for drinking and farming for more than 500 million people. The west coast of North America, which gets much of its water from glaciers in mountain ranges such as the Rocky Mountains and Sierra Nevada, also would be affected. Glaciers aren't the only worry that the developing nations have, sea level is also reported to rise as climate change progresses, reducing the amount of land available for agriculture.

In other parts of the world, a big effect will be low yields of grain according to the World Food Trade Model, specifically in the low latitude regions where much of the developing world is located. From this the price of grain will rise, along with the developing nations trying to grow the grain. Due to this, every 2-2.5 per cent price hike will increase the number of hungry people by 1 per cent. Low crop yields are just one of the problem facing farmers in the low latitudes and tropical regions. The timing and length of the growing seasons, when farmers plant their crops, are going to be changing dramatically, per the USDA, due to unknown changes in soil temperature and moisture conditions.

Food Justice

An alternative view takes a collective approach to achieve food security. It notes that globally enough food is produced to feed the entire world population at a level adequate to ensure that everyone can be free of hunger and fear of starvation. That no one should live without enough food because of economic constraints or social inequalities is the basic goal.

This approach is often referred to as food justice and views food security as a basic human right. It advocates fairer distribution of food, particularly grain crops, as a means of ending chronic hunger and malnutrition. The core of the Food Justice movement is the belief that what is lacking is not food, but the political will to fairly distribute food regardless of the recipient's ability to pay.

Food Sovereignty

A third approach is known as food sovereignty; though it overlaps with food justice on several points, the two are not identical. It views the business practices of multinational corporations as a form of neocolonialism. It contends that multinational corporations have the financial resources available to buy up the agricultural resources of impoverished nations, particularly in the tropics. They also have the political clout to convert these resources to the exclusive production of cash crops for sale to industrialised nations outside of the tropics, and in the process to squeeze the poor off of the more productive lands.

Under this view subsistence farmers are left to cultivate only lands that are so marginal in terms of productivity as to be of no interest to the multinational corporations. Likewise, food sovereignty holds it to be true that communities should be able to define their own means of production and that food is a basic human right.

With several multinational corporations now pushing agricultural technologies on developing countries, technologies that include improved seeds, chemical fertilizers, and pesticides, crop production has become an increasingly analysed and debated issue. Many communities calling for food sovereignty are protesting the imposition of Western technologies on to their indigenous systems and agency.

Those who hold a "food sovereignty" position advocate banning the production of most cash crops in developing nations, thereby leaving the local farmers to concentrate on subsistence agriculture. In addition, they oppose allowing low-cost subsidised food from industrialised nations into developing countries, what is referred to as "import dumping". Import dumping also happens by way of food aid distribution through programmes like the USA's "Food for Peace" initiative.

World Food Summit

The World Food Summit was held in Rome in 1996, with the aim of renewing global commitment to the fight against hunger. The Food and Agriculture Organisation of the United Nations (FAO) called the summit in response to widespread under-nutrition and growing concern about the capacity of agriculture to meet future food needs. The conference produced two key documents, the Rome Declaration on World Food Security and the World Food Summit Plan of Action.

The Rome Declaration calls for the members of the United Nations to work to halve the number of chronically undernourished people on the Earth by the year 2015. The Plan of Action sets a number of targets for government and non-governmental organisations for achieving food security, at the individual, household, national, regional and global levels.

World Summit on Food Security

The World Summit on Food Security took place in Rome, Italy, between 16 and 18 November 2009. The decision to convene the summit was taken by the Council of FAO in June 2009, at the proposal of FAO Director-General Dr Jacques Diouf. Heads of State and Government attended the summit, which took place at the FAO's headquarters.

Role of World Bank

1. Global Food Security Programme: Launched in April 2010, six countries and Bill and Melinda Foundations has pledge $ 925 mn for food security. Till date it has helped 8 countries. Promote agriculture, research, trade in agriculture etc.
2. Also Launched Global Food Crisis Response Programme: Given grants to ~40nations for seeds, etc. for improving productivity.
3. Also in process of increasing its yearly spending for agriculture to $6-8 bn from earlier 4bn

4. Also runs several nutrition programme across the world *e.g.*, vitamin A doses for children, school meals atc.

Achieving Food Security

"The number of people without enough food to eat on a regular basis remains stubbornly high, at over 800 million, and is not falling significantly. Over 60 per cent of the world's undernourished people live in Asia, and a quarter in Africa. The proportion of people who are hungry, however, is greater in Africa (33%) than Asia (16%). The latest FAO figures indicate that there are 22 countries, 16 of which are in Africa, in which the undernourishment prevalence rate is over 35 per cent."

By way of comparison, in one of the largest food producing countries in the world, the United States, approximately one out of six people are "food insecure", including 17 million children, according to the U.S. Department of Agriculture. Food insecurity is measured in the United States by questions in the Census Bureau's Current Population Survey. The questions asked are about anxiety that the household budget is inadequate to buy enough food, inadequacy in the quantity or quality of food eaten by adults and children in the household, and instances of reduced food intake or consequences of reduced food intake for adults and for children. A National Academy of Sciences study commissioned by the USDA criticised this measurement and the relationship of "food security" to hunger, adding "it is not clear whether hunger is appropriately identified as the extreme end of the food security scale." In its "The State of Food Insecurity in the World 2003", FAO states that:

'In general the countries that succeeded in reducing hunger were characterised by more rapid economic growth and specifically more rapid growth in their agricultural sectors. They also exhibited slower population growth, lower levels of HIV and higher ranking in the Human Development Index'.

As such, according to FAO, addressing agriculture and population growth is vital to achieving food security. Other organisations and people have come to this same conclusion, and advocate improvements in agriculture and population control.

USAID proposes several key steps to increasing agricultural productivity which is in turn key to increasing rural income and reducing food insecurity. They include:

- Boosting agricultural science and technology. Current agricultural yields are insufficient to feed the growing populations. Eventually, the rising agricultural productivity drives economic growth.
- Securing property rights and access to finance.
- Enhancing human capital through education and improved health.
- Conflict prevention and resolution mechanisms and democracy and governance based on principles of accountability and transparency

in public institutions and the rule of law are basic to reducing vulnerable members of society.

The UN Millennium Development Goals are one of the initiatives aimed at achieving food security in the world. In its list of goals, the first Millennium Development Goal states that the UN "is to eradicate extreme hunger and poverty", and that "agricultural productivity is likely to play a key role in this if it is to be reached on time".

"Of the eight Millennium Development Goals, eradicating extreme hunger and poverty depends on agriculture the most. (MDG 1 calls for halving hunger and poverty by 2015 in relation to 1990.)

Notably, the gathering of wild food plants appears to be an efficient alternative method of subsistence in tropical countries, which may play a role in poverty alleviation.

The Agriculture-hunger-poverty Nexus

Eradicating hunger and poverty requires an understanding of the ways in which these two injustices interconnect. Hunger, and the malnourishment that accompanies it, prevents poor people from escaping poverty because it diminishes their ability to learn, work, and care for themselves and their family members.

Food insecurity exists when people are undernourished as a result of the physical unavailability of food, their lack of social or economic access to adequate food, and/or inadequate food utilisation. Food-insecure people are those individuals whose food intake falls below their minimum calorie (energy) requirements, as well as those who exhibit physical symptoms caused by energy and nutrient deficiencies resulting from an inadequate or unbalanced diet or from the body's inability to use food effectively because of infection or disease. An alternative view would define the concept of food insecurity as referring only to the consequence of inadequate consumption of nutritious food, considering the physiological utilisation of food by the body as being within the domain of nutrition and health.

Malnourishment also leads to poor health hence individuals fail to provide for their families. If left unaddressed, hunger sets in motion an array of outcomes that perpetuate malnutrition, reduce the ability of adults to work and to give birth to healthy children, and erode children's ability to learn and lead productive, healthy, and happy lives. This truncation of human development undermines a country's potential for economic development–for generations to come.

There are strong, direct relationships between agricultural productivity, hunger, and poverty. Three-quarters of the world's poor live in rural areas and make their living from agriculture. Hunger and child malnutrition are greater in these areas than in urban areas. Moreover, the higher the proportion of the rural population that obtains its income solely from subsistence farming (without the benefit of pro-poor technologies and access to markets), the higher the incidence of malnutrition. Therefore, improvements in agricultural productivity aimed at small-scale farmers will benefit the rural poor first.

Increased agricultural productivity enables farmers to grow more food, which translates into better diets and, under market conditions that offer a level playing field, into higher farm incomes. With more money, farmers are more likely to diversify production and grow higher-value crops, benefiting not only themselves but the economy as a whole."

Researchers suggest forming an alliance between the emergency food programme and CSA Farms, as currently food stamps cannot be used at farmer's markets and places in which food is less processed and grown locally.

Children

On 2008-04-29, a UNICEF UK report found that the world's poorest and most vulnerable children are being hit the hardest by the impact of climate change. The report, "Our Climate, Our Children, Our Responsibility: The Implications of Climate Change for the World's Children," says access to clean water and food supplies will become more difficult, particularly in Africa and Asia.

Wheat Stem Rust

An epidemic of stem rust on wheat caused by race Ug99 is currently spreading across Africa and into Asia and is causing major concern. A virulent wheat disease could destroy most of the world's main wheat crops, leaving millions to starve. The fungus has spread from Africa to Iran, and may already be in Pakistan.

The genetic diversity of the crop wild relatives of wheat can be used to improve modern varieties to be more resistant to rust. In their centers of origin wild wheat plants are screened for resistance to rust, then their genetic information is analysed and finally wild plants and modern varieties are crossed through means of modern plant breeding in order to transfer the resistance genes from the wild plants to the modern varieties.

Dictatorship and Kleptocracy

Nobel Prize-winning economist Amartya Sen has observed that "there is no such thing as an apolitical food problem." While drought and other naturally occurring events may trigger famine conditions, it is government action or inaction that determines its severity, and often even whether or not a famine will occur. The 20th century is full of examples of governments undermining the food security of their own nations–sometimes intentionally.

When governments come to power by force or rigged elections, and not by way of fair and open elections, their base of support is often narrow and built upon cronyism and patronage. Under such conditions "The distribution of food within a country is a political issue. Governments in most countries give priority to urban areas, since that is where the most influential and powerful families and enterprises are usually located.

The government often neglects subsistence farmers and rural areas in general. The more remote and underdeveloped the area the less likely the government will be to effectively meet its needs. Many agrarian policies, especially the pricing of agricultural commodities, discriminate against rural areas.

Governments often keep prices of basic grains at such artificially low levels that subsistence producers cannot accumulate enough capital to make investments to improve their production.

Thus, they are effectively prevented from getting out of their precarious situation." Further dictators and warlords have used food as a political weapon, rewarding their supporters while denying food supplies to areas that oppose their rule. Under such conditions food becomes a currency with which to buy support and famine becomes an effective weapon to be used against the opposition.

Governments with strong tendencies towards kleptocracy can undermine food security even when harvests are good. When government monopolises trade, farmers may find that they are free to grow cash crops for export, but under penalty of law only able to sell their crops to government buyers at prices far below the world market price. The government then is free to sell their crop on the world market at full price, pocketing the difference. This creates an artificial "poverty trap" from which even the most hard working and motivated farmers may not escape.

When the rule of law is absent, or private property is non-existent, farmers have little incentive to improve their productivity. If a farm becomes noticeably more productive than neighbouring farms, it may become the target of individuals well connected to the government. Rather than risk being noticed and possibly losing their land, farmers may be content with the perceived safety of mediocrity.

As pointed out by William Bernstein in his book *The Birth of Plenty*: "Individuals without property are susceptible to starvation, and it is much easier to bend the fearful and hungry to the will of the state. If a [farmer's] property can be arbitrarily threatened by the state, that power will inevitably be employed to intimidate those with divergent political and religious opinions."

Economic Approaches

There are many economic approaches advocated to improve food security in developing countries. Three typical approaches are listed below. The first is typical of what is advocated by most governments and international agencies. The other two are more common to non-governmental organisations (NGO's).

Biotechnology for Smallholders in the (sub) Tropics

The area sown to genetically engineered crops in developing countries is rapidly catching-up with the area sown in industrial nations. According to the International Service for the Acquisition of Agri-biotech Applications (ISAAA), genetically engineered (biotech, GM) crops were grown by approximately 8.5 million farmers in 21 countries in 2005, up from 8.25 million farmers in 17 countries in 2004. The largest increase in biotech crop area in any country in 2005 was in Brazil, provisionally estimated at 44,000 km² (94,000 km² in 2005 compared with 50,000 km² in 2004). India had by far the largest year-on-year proportional increase, with almost a threefold increase from 5,000 km² in 2004 to 13,000 km² in 2005. However, it should be noted that the ISAAA is funded by organisations including prominent agricultural biotechnology corporations, such as Monsanto and Bayer, and there have been several challenges made to the accuracy of ISAAA's global figures.

Current high regulatory costs imposed on varieties created by the more modern methods are a significant hurdle for development of genetically engineered crops well suited to developing country farmers by modern genetic methods. Once a new variety is developed, however, seed provides a good vehicle for distribution of improvements in a package that is familiar to the farmer. Currently there are some institutes and research groups that have projects in which biotechnology is shared with contact people in less-developed countries on a non-profit basis. These institutes make use of biotechnological methods that do not involve high research and registration costs, such as conservation and multiplication of germplasm and phytosanitation.

Apart from genetic engineering, other forms of biotechnology also hold promise for enhancing food security. For instance, perennial rice is being developed in China, which could dramatically reduce the risk of soil erosion on upland smallholder farms.

Westernised View

Conventional thinking in westernised countries is that maximising farmers' profits is the surest way of maximising agricultural production; the higher a farmer's profit, the greater the effort that will be forthcoming, and the greater the risk the farmer is willing to take.

Place into the hands of farmers the largest number and highest quality tools possible (tools is used here to refer to improved production techniques, improved seeds, secure land tenure, accurate weather forecasts, etc.) However, it is left to the individual farmer to pick and choose which tools to use, and how to use them, as farmers have intimate knowledge of their own land and local conditions.

As with other businesses, a percentage of the profits are normally reinvested into the business in the hopes of increasing production, and hence increase future profits. Normally higher profits translate into higher spending on technologies designed to boost production, such as drip irrigation systems, agriculture education, and greenhouses. An increased profit also increases the farmer's incentive to engage in double-cropping, soil improvement programmes, and expanding usable area.

Fossil Fuel Dependence

While agricultural output increased as a result of the Green Revolution, the energy input into the process (that is, the energy that must be expended to produce a crop) has also increased at a greater rate, so that the ratio of crops produced to energy input has decreased over time. Green Revolution techniques also heavily rely on chemical fertilizers, pesticides and herbicides, some of which must be developed from fossil fuels, making agriculture increasingly reliant on petroleum products.

Between 1950 and 1984, as the Green Revolution transformed agriculture around the globe, world grain production increased by 250 per cent. The energy for the Green Revolution was provided by fossil fuels in the form of fertilizers (natural gas), pesticides (oil), and hydrocarbon fueled irrigation.

The authors of this study believe that the mentioned agricultural crisis will only begin to impact us after 2020, and will not become critical until 2050. The oncoming peaking of global oil production (and subsequent decline of production), along with the peak of North American natural gas production will very likely precipitate this agricultural crisis much sooner than expected. Geologist Dale Allen Pfeiffer claims that coming decades could see spiraling food prices without relief and massive starvation on a global level such as never experienced before.

However, one should take note that, (numbers taken from the CIA World Factbook), the country of Bangladesh achieved food self-sufficiency in 2002 with both a far higher population density than the USA (~1000 inhabitants per square kilometre in comparison to just 30/km^2 for the USA - so this is more than 30 times as many), and at only a tiny fraction of the USA's usage of oil, gas, and electricity. Also, pre-industrial Chinese mini-farmers/gardeners developed techniques to feed a population of more than 1000 people per square kilometre. Hence, the dominant problem is not energy availability but the need to stop and revert soil degradation.

Genetic Engineering and Loss of Biodiversity

In agriculture and animal husbandry, the Green Revolution popularised the use of conventional hybridisation to increase yield by creating "high-yielding varieties". Often the handful of hybridised breeds originated in developed countries and were further hybridised with local varieties in the rest of the developing world to create high yield strains resistant to local climate and diseases. Local governments and industry have been pushing hybridisation which has resulted in several of the indigenous breeds becoming extinct or threatened. Disuse because of unprofitability and uncontrolled intentional and unintentional cross-pollination and crossbreeding (genetic pollution), formerly huge gene pools of various wild and indigenous breeds have collapsed causing widespread genetic erosion and genetic pollution. This has resulted in loss of genetic diversity and biodiversity as a whole.

A genetically modified organism (GMO) is an organism whose genetic material has been altered using the genetic engineering techniques generally known as recombinant DNA technology. Genetically Modified (GM) crops today have become a common source for genetic pollution, not only of wild varieties but also of other domesticated varieties derived from relatively natural hybridisation.

Genetic erosion coupled with genetic pollution may be destroying unique genotypes, thereby creating a hidden crisis which could result in a severe threat to our food security. Diverse genetic material could cease to exist which would impact our ability to further hybridise food crops and livestock against more resistant diseases and climatic changes.

Agricultural and Livestock Biodiversity

Genetic erosion in agricultural and livestock biodiversity is the loss of genetic diversity, including the loss of individual genes, and the loss of particular combinants of genes (or gene complexes) such as those manifested in locally adapted landraces of domesticated animals or plants adapted to the natural environment in which they originated. The term genetic erosion is sometimes used in a narrow sense, such as for the loss of alleles or genes, as well as more broadly, referring to the loss of varieties or even species. The major driving forces behind genetic erosion in crops are: variety replacement, land clearing, overexploitation of species, population pressure, environmental degradation, overgrasing, policy and changing agricultural systems.

The main factor, however, is the replacement of local varieties of domestic plants and animals by high yielding or exotic varieties or species. A large number of varieties can also often be dramatically reduced when commercial varieties (including GMOs) are introduced into traditional farming systems. Many researchers believe that the main problem related to agro-ecosystem management is the general tendency towards genetic and ecological uniformity imposed by the development of modern agriculture.

Intellectual Property Rights

There is much debate on whether IPRs hurt or harm independent development in terms or agriculture and food production. Hartmut Meyer and Annette von Lossau describe both sides of the issue, while saying "Among scholars, the thesis that the impetus to self-determined development and the protection of intellectual property go hand in hand is disputed - to put it mildly. Many studies have concluded that there is virtually no positive correlation between establishing self-sustained economic growth and ensuring protection of intellectual property rights.

CAUSES OF CLIMATE CHANGES

Climate changes in response to changes in the global energy balance. On the broadest scale, the rate at which energy is received from the sun and the rate at which it is lost to space determine the equilibrium temperature and climate of Earth. This energy is then distributed around the globe by winds, ocean currents, and other mechanisms to affect the climates of different regions.

Factors that can shape climate are called climate forcings or "forcing mechanisms".

These include such processes as variations in solar radiation, deviations in the Earth's orbit, mountain-building and continental drift, and changes in greenhouse gas concentrations.

There are a variety of climate change feedbacks that can either amplify or diminish the initial forcing. Some parts of the climate system, such as the oceans and ice caps, respond slowly in reaction to climate forcings, while others respond more quickly. Forcing mechanisms can be either "internal" or "external". Internal forcing mechanisms are natural processes within the climate system itself (*e.g.*, the meridional overturning circulation). External forcing mechanisms can be either natural (*e.g.*, changes in solar output) or anthropogenic (*e.g.*, increased emissions of greenhouse gases).

Whether the initial forcing mechanism is internal or external, the response of the climate system might be fast (*e.g.*, a sudden cooling due to airborne volcanic ash reflecting sunlight), slow (*e.g.* thermal expansion of warming ocean water), or a combination (*e.g.*, sudden loss of albedo in the arctic ocean as sea ice melts, followed by more gradual thermal expansion of the water). Therefore, the climate system can respond abruptly, but the full response to forcing mechanisms might not be fully developed for centuries or even longer.

Terminology

The most general definition of *climate change* is a change in the statistical properties of the climate system when considered over long periods of time, regardless of cause. Accordingly, fluctuations over periods shorter than a few decades, such as El Nino, do not represent climate change.

The term sometimes is used to refer specifically to climate change caused by human activity, as opposed to changes in climate that may have resulted as part of Earth's natural processes. In this latter sense, used especially in the context of environmental policy, the term *climate change* today is synonymous with anthropogenic global warming. Within scientific journals, however, *global warming* refers to surface temperature increases, while *climate change* includes global warming and everything else that increasing greenhouse gas amounts will affect.

Internal Forcing Mechanisms

Natural changes in the components of earth's climate system and their interactions are the cause of internal climate variability, or "internal forcings." Scientists generally define the five components of earth's climate system to include Atmosphere, hydrosphere, cryosphere, lithosphere (restricted to the surface soils, rocks, and sediments), and biosphere.

Ocean Variability

The ocean is a fundamental part of the climate system, some changes in it occurring at longer timescales than in the atmosphere, massing hundreds of times more and having very high thermal inertia (such as the ocean depths still lagging today in temperature adjustment from the Little Ice Age).

Short-term fluctuations (years to a few decades) such as the El Nino-Southern Oscillation, the Pacific decadal oscillation, the North Atlantic oscillation, and the Arctic oscillation, represent climate variability rather than climate change. On longer time scales, alterations to ocean processes such as thermohaline circulation play a key role in redistributing heat by carrying out a very slow and extremely deep movement of water, and the long-term redistribution of heat in the world's oceans.

EFFECTS OF SYSTEMATIC CLIMATE VARIABILITY ON EPIDEMICS

Another significant climate trend that has been linked to systematic changes in temperature and precipitation is an increase in the variability, or extremes, of climate. This change in extremes can alter not only the intensity of individual events, such as storms and floods, but the timing and spatial patterns of weather as well. Since the mid-1800s the average surface temperature of the globe has risen about 0.4°C-0.6°C, and periods of persistent warming can, in general, be associated with increased variability. The IPCC projects that the warming trend may be accompanied by more intense heat waves and altered drought and rainfall patterns.

As reported in the first issue of Consequences, data from the National Climatic Data Centre—this nation's main repository for meteorological data—indicate that, since the 1970s, extreme weather events have indeed increased in the continental U.S. On average, periods of drought are systematically longer, and bursts of precipitation (greater than two inches of rain over twenty-four hours) are more frequent. A warmed atmosphere accelerates evaporation, and for every rise of 1°C, the air can hold 6 percent more water. One consequence is that we are now receiving a greater percentage of precipitation in the form of sudden, intense bursts that are more typical, for example, in the tropics. Longer droughts and more heavy bursts of rain, accompanied by flash floods, were more common in the 1980s than in the 1970s, and more so in the 1990s than in the 1980s.

Extreme events—floods, storms, droughts, and uncontained fires—can be devastating for agriculture, for human settlements, and for health. More, brief cold snaps are also possible, and winter storms, like heat waves, bring an increase in cardiac deaths. Floods spread bacteria, viruses, and chemical contaminants, foster the growth of fungi, and contribute to the breeding of insects. Prolonged droughts, interrupted by heavy rains, favour population explosions of both insects and rodents. Extreme weather events (often associated with the recurring climatic conditions that are initiated by large-scale changes in sea surface temperatures in the Pacific, known as El Nino/La Nina events) have been accompanied by new appearances of harmful algal blooms in Asia and North America, and—in Latin America and Asia—by outbreaks of malaria and various water-borne diseases, such as typhoid, hepatitis A, bacillary dysentery, and cholera.

Rodents: Synergies and Surprises

Rodents are today a growing problem in the U.S., Latin America, Africa, Europe, Asia, and Australia. These preeminent opportunists are believed to be the fastest reproducing mammal and they eat everything humans do, thrive on contaminated water and food, and are extremely capable swimmers. Meadow voles, whose numbers are kept in check by predatory marsh hawks, can have up to seventeen broods a year, for example, each of half a dozen offspring. Rodents consume 20 percent of the world's grain, including almost a seventh in our own country, and up to three quarters of what is grown and stored in some African nations. Rodents can also carry diseases. A controlled experiment with Canadian snowshoe hares depicts how multiple factors can act synergistically to greatly increase the number of rodents. Excluding predators by confining the rabbits to cages led to a population doubling, compared to the fate of a number of similar animals in the wild. Augmenting food tripled hare density. Together the interventions in the controlled experiment resulted in more than a ten-fold population explosion.

Rats, Mice, and the Hantavirus

The story of the hantavirus illustrates a similar synergy in the case of microbial agents. A prolonged drought in the U.S. southwest in the early 1990s reduced the populations of animals such as owls, coyotes, and snakes that prey on rodents. When the drought yielded to intense rains in 1993, the grasshoppers and pinon nuts on which rodents feed became more abundant. The result, when combined with the drop in predators, was a ten-fold increase in rats and mice by June of that year. An outcome was the emergence of a "new" disease called Hantavirus Pulmonary Syndrome: from a virus—perhaps already present, but dormant—transmitted through rodent saliva, urine, and droppings. Predators returned, however, and by summer's end the outbreak had abated. Rodent-borne hantaviruses have resurged in several European nations, and most notably in the former Soviet Union and in Yugoslavia. In late 1996, hantavirus infection emerged in South America in western Argentina. Within the first few months at least ten deaths resulted, frightening off tourists and threatening the economic livelihood of the region. Hantavirus Pulmonary Syndrome has now appeared in Bolivia, Brazil, Canada, Chile, Ecuador, Paraguay, and Uruguay, with about 50 percent mortality and several cases of person-to-person transmission.

Disease Clusters

In August 1995 the eastern, tropical region of the Pacific Ocean surface turned cold, initiating a La Nina event that would last until late 1996. Along the Caribbean coast of Colombia a summer 1995 heat wave was followed by the heaviest August rainfall in fifty years, ending a long drought that accompanied the preceding, prolonged El Nino conditions of 1990-1995. The heat and flooding precipitated a cluster of diseases involving mosquitoes (Venezuelan equine encephalitis and dengue fever), rodents (leptospirosis), and toxic algae (that killed 350 tons of fish in their largest coastal lagoon). Prolonged anomalous conditions of the sort that applied in 1990-1995 can also have cumulative biological consequences. In New Orleans, for example, five years without a killing frost was associated with an explosion of mosquitoes, termites, and cockroaches. Termites persisted inside trees into the cold winter of 1995-1996, and now threaten to destroy stands of New Orleans' fabled "mighty oaks." We may have only just begun to understand the true biological impacts of the persistent anomalous climate of the 1990s.

Some other Examples

Evidence of another rodent-borne disease —leptospirosis— is increasingly reported in U.S. urban centres, in areas where the disposal of sewage and other measures to protect public health have declined. In 1995, there were substantial outbreaks of the disease in Central America and Colombia, as heavy (La Nina) rains, following the prolonged El Nino drought, drove rodents scurrying from their burrows. Leptospirosis is treatable with antibiotics; but there were fatalities in 1995 before the diagnosis was established.

A combination of stresses contributed to the sudden appearances of several rodent-borne viral hemorrhagic fevers in rural Latin America in the past several decades: Junin in Argentina (1953), machupo in Bolivia (1962 and 1996), guaranito in Venezuela, and sabia in Brazil. In Bolivia, systematic clearing of trees apparently shifted populations of a variety of disease-carrying mice, known as calomys, from forest to field settings where they became dominant. Heavy applications of DDT —meant to eradicate malaria— helped reduce their natural predators. When cats were reintroduced to the area in 1962, the epidemic of Bolivian hemorrhagic fever was abated, although not until it had killed 10 to 20 percent of the inhabitants of the small villages in which the disease was present. Habitat loss and excessive use of toxic chemicals acted together in this case. In southern Africa, rodent populations exploded as a consequence of climate variability, when heavy and then lighter rains came in 1993 and 1994, on the heels of six years of prolonged drought. When the rains came, rodents found themselves in a world where avian and land predators were virtually absent. Moreover, because so many draft animals had also succumbed, there was little tillage of the land and the underground burrows in which rodents live went largely undisturbed. After an initial successful harvest in 1993, the maize crop in Zimbabwe was decimated by rodents. Soon after, human plague broke out in Zimbabwe and on the borders of neighbouring Malawi and Mozambique, carried—as in the devastating plague of 14th century Europe—by fleas on rats. Subsequently, a rodent-borne virus took the lives of eighty-one elephants in South Africa's Kruger Park, and plague returned in the summer of 1997.

Plague in India resurfaced in 1994, following a blistering summer when temperatures reached 124°F, leaving animals prostrate in the northern part of the country and fueling the breeding of fleas in houses that held stored grain. The unusually heavy monsoons following the 1994 heat wave led to population crowding in Surat on the western coast, north of Bombay, and an apparent outbreak of pneumonic (person-to-person) plague. Cases of malaria and dengue fever also surged in the wake of flooding. Meanwhile, in Australia, rodents emerged as serious crop pests in 1995, accompanying the same prolonged El Nino of the early 1990s and the ensuing years of intense drought.

Current land-use practices and the overuse of chemicals to control pests may increase the chances for such "nasty synergies." Climate variability is also a key element in upsurges of pests—and were climate to become more unstable, it could exert an even greater influence on ecological dynamics and the patterns of infectious diseases in the future. A disturbance in one factor can be destabilizing; but multiple perturbations can reduce the resistance and the resilience of a entire system.

2

Responses of Forests to Variation in Climate Change

The absolute limits on tree distribution are generally associated with climatic limitations. There has therefore been an effort to define these limits, as well as the optimum conditions of temperature and moisture for broadly representative types of forests.

DETERMINATION OF TEMPERATURE OPTIMUM AND LIMITS

Ecophysiological studies have established the lower temperature limits for tropical, subtropical, temperate and boreal forests. Many subtropical species are able to concentrate solute and thus lower the freezing point in cells below that tolerated by tropical trees. Some temperate trees extend the freezing threshold down to -45°C via the mechanism of supercooling, whereas boreal species are able to transfer water out of the nucleus into cell walls and thereby withstand exposure to liquid nitrogen at -196°C. These adaptations to low temperatures serve to define the potential geographical limits of various types of forests.

Defining the optimum temperature for growth or photosynthesis has traditionally been made in the laboratory on seedlings or from excised branches or buds of mature trees. By measuring the production rates of metabolic heat and CO_2 efflux of excised buds at varying temperatures, Anekonda *et al.* (1993, 1994) described relationships with growth and geographical origins of *Sequoia sempervirens*.

However, the temperature optimum for photosynthesis often exhibits seasonal adjustment. Intuitively, we would think that optimal growth for a species would occur at temperatures typical of those that occur in the species' native range. In fact, this is not true. Studies by Hawkins and Sweet (1989) demonstrated that optimum temperatures for a broad array of tree species native to New Zealand averaged nearly 8°C above typical midsummer temperatures, whereas faster-growing species introduced from North America were closely adapted to the present New Zealand climate. Further refinements in growth-room experiments identified the importance of root temperature. Root temperature affects the rate that water may be taken up through root membranes, which helps explain the distribution patterns of closely related genera across broad climatic zones.

PHYSIOLOGICAL DEFINITION OF SOIL DROUGHT

Inadequate precipitation can lead to a soil water deficit. Ralph Slatyer (1967) set forth the basic principles of how plants respond to soil drought but until Per Scholander and colleagues at Scripps Institution of Oceanography constructed a portable pressure chamber, ecophysiologists were unable to measure water potential gradients in tall trees. With the pressure chamber, relationships between relative water content, solute potential and incipient plas-molysis could be assessed and seasonal comparisons made to evaluate the physiological impact of drought. In drought-prone areas, seedlings often experience near-lethal water potentials because their roots do not extend as deeply as those of mature trees.

The pressure chamber provided the first direct measure of water availability to deeply rooted trees when measurements were taken at night under conditions when transpiration approaches zero. The fraction of water extracted from different depths within the soil profile has more recently been determined through analysis of the isotopic composition ($8^{18}O$ and $8^{2}H$) of xylem sap.

Sapwood as a Water Reservoir in Trees

Trees contain a large volume of sapwood, which is significant because this tissue serves during the day and through periods of drought as a temporary reservoir of water that extends the opportunities for trees to photosynthesize. The extraction of large amounts of water from stems and branches involves some cavitation in the water-conducting elements of the xylem. Rapid recharge of the sapwood reservoir occurs through uptake of water by roots and through foliage and twigs.

The amount of water released from sapwood for a specific increase in the water potential gradient defines the 'capacitance', an essential term in modelling the water balance of trees and forests. As more water is extracted from the sapwood, the water transport pathway between the soil and evaporating surfaces of the foliage becomes less efficient, with implications for both transpiration and photosynthesis.

Transpiration by Forests

Transpiration was the first physiological process directly measured for an entire forest by employing an energy budget analysis. A forest, unlike vegetation of shorter more uniform stature, represents an aero-dynamically rough surface, a property that creates turbulence in the flow of air, even under calm conditions. As a consequence of this aerodynamic roughness, forests can absorb large amounts of solar radiation without experiencing temperatures that would be lethal to shorter vegetation, even when transpiration is restricted.

Stomatal response is closely linked to the pho-tosynthetic process. Consequently, accurate transpiration models separate the photosynthetically active component of solar radiation (400-700nm) from the total and take into account hydraulic constraints and variation in ambient concentrations of CO_2.

Most models that predict the mean canopy conductance of whole forests are still empirical, developed from responses recorded under a range of environmental conditions using gas-exchange cuvettes or inferred from flux measurements determined when the forest canopy is dry.

From the extensive literature published on leaf and canopy stomatal conductance, some important generalizations have emerged. First, the maximum canopy conductance for a wide variety of forests averages two to four times the maximum value measured on individual leaves. The leaf area of woodlands and forests is expressed in terms of leaf area index (m^2 of projected leaf surface per m^2 of ground) and ranges from less than $1 m^2 m^{-2}$ to $12 m^2 m^{-2}$.

The maximum canopy conductance for forests is lower than the maximum that would be expected from the maximum canopy conductance of individual leaves because the illumination of individual leaves decreases as they shade each other when the total canopy leaf area increases. This reduces the stomatal conductance of individual leaves below that which would occur in full sunlight. As a result of self-shading, the maximum canopy conductance approaches a constant at leaf area indices greater than $3.0\ m^2\ m^{-2}$. These two generalities greatly simplify the estimation of maximum stomatal conductance and reduce the need for accurate assessment of leaf area indices above $3.0 m^2 m^{-2}$ for purposes of estimating transpiration.

Although the *maximum* canopy conductance may not differ significantly among forests, this does not mean that reliable estimates of *actual* canopy stomatal conductance can be extrapolated from measurements made on seedlings or saplings to tall trees. Trees have deeper root systems and a larger volume of sapwood than seedlings or saplings, yet even with these benefits mature trees offer an increasingly long and tortuous path for water movement from roots to leaves.

As a result, comparative studies between young and older trees often show that when water is readily available in the soil, older trees may be more sensitive to atmospheric vapour pressure deficits. The differences are not simply in response to height and the hydrostatic gradient of $0.01\ MPam^{-1}$. The spacing between individual trees also plays a role. As trees increase in size, branches elongate as a function of the space between trees. Longer branches offer a less efficient pathway for water transport than the main stem, and can reduce transpiration and photosynthesis per unit of leaf area by hastening stomatal closure during the day.

OBSERVING CLIMATE VARIABILITY AND CHANGE

The Earth's climate is dynamic and naturally varies on seasonal, decadal, centennial, and longer timescales. Each "up and down" fluctuation can lead to conditions which are warmer or colder, wetter or drier, more stormy or quiescent. Analyses of decadal and longer climate records and studies based on climate models suggest that many changes in recent decades can be attributed to human actions; these decadal trends are referred to as climate change. The effects of climate variability and change ripple throughout the environment and society - indeed touching nearly all aspects of the human endeavor and the environment. These factors underlie NOAA's mission to observe, understand, and predict climate variability and change.

VARIATIONS AND CHANGES OCCURRING IN EARTH'S CLIMATE

Variability. Perhaps the most well understood occurrence of climate variability is the naturally occurring phenomenon known as the El Niño-Southern Oscillation (ENSO), an interaction between the ocean and the atmosphere over the tropical Pacific Ocean that has important consequences for weather around the globe. The ENSO cycle is characterized by coherent and strong variations in sea-surface temperatures, rainfall, air pressure, and atmospheric circulation across the equatorial Pacific. El Niño refers to the warm phase of the cycle, in which above-average sea-surface temperatures develop across the east-central tropical Pacific.

La Niña is the cold phase of the ENSO cycle. The swings of the ENSO cycle typically occur on a time scale of a few years. These changes in tropical rainfall affect weather patterns throughout the world. Because of the importance of ENSO, NOAA has established a special ocean - atmosphere observing system in the tropical Pacific; this enables forecasts of El Niño to be made several seasons in advance.

Climate variability is manifested in other ways as well. Decadal and seasonal shifts in wind patterns and sea surface temperatures in the Atlantic cause changes in hurricane frequency, for example. Sometimes climate varies in ways that are random or not fully explainable. The Dust Bowl of the 1930s in the United States is one such example. Because of the social and economic importance of understanding such climate fluctuations, NOAA routinely monitors past and current climate. Many of the longer period fluctuations are linked to the ocean. NOAA with its international partners is implementing a new global ocean observing system. This will lead to new understanding which will improve our predictions of climate variability and change.

Change. The amount of energy entering and escaping from Earth is the determining factor in climate. Any changes to that balance-either the input or the output-will cause a directional change in climate. Observations have conclusively demonstrated that the atmospheric abundance of greenhouse gases has risen dramatically since the onset of the Industrial Age. Human activities such as fossil fuel burning and deforestation have caused this increase. The atmospheric concentration of carbon dioxide (CO2), has risen by 31% since 1750 and is now at the highest concentration seen in the last 420,000 years (and likely higher than any concentration seen for the last 20 million years). Other greenhouse gases that have increased since 1750 are methane (up by 151%), nitrous oxide (up by 17%), halocarbons (rising rapidly since 1950 but slowing or decreasing in recent years because of international agreements to protect the ozone layer), and tropospheric ozone (up by 36%). These gases absorb heat that would otherwise escape to space.

The natural greenhouse effect warms the surface of the planet to temperatures that are hospitable for life. Indeed, without it we would have a frozen planet with surface temperatures of 0°F (-18°C), some 59°F colder than the current value. OAR has taken the lead internationally to build and maintain a system to monitor greenhouse gases and atmospheric fine particles (such as soot and sulfate aerosols) that influence climate, as well as other important trace gases.

The energy *input* to Earth from the Sun can vary, for example because of changes in Earth's orbital path, changes in the tilt angle of the Earth, or variations in the energy coming from the Sun. These changes occur over long time scales and lead to the cycling of Ice Ages and interglacial periods that have occurred over the last 500,000 years. An additional factor which can cause climatic fluctuations on yearly to decadal timescales are volcanoes. Volcanic eruptions lead to enhancements of stratospheric and tropospheric aerosols which for the most part reflect solar radiation, hence leading to global cooling on a global average. However, regionally these can lead to warmer conditions by changing atmospheric circulation patterns, especially in the wintertime.

Global observations of climate-related quantities have made it possible to document and analyze the behavior of Earth's climate. An international scientific body known as the Intergovernmental Panel on Climate Change (IPCC) periodically updates the state of scientific understanding with regard to climate science. Hundreds of scientists worldwide participate in the preparation and review of these reports. The latest such IPCC report is *Climate Change 2001: The Scientific Basis*. Also in 2001, the U.S. National Research Council issued a report on *Climate Change Science: An Analysis of Some Key Questions*. Some of the findings of the IPCC report and the NRC report with regard to the occurrence of climate change are:

- On the global average, the surface air temperature of the Earth has increased by about 1.0 ± 0.4°F (0.6 ± 0.2°C) since the late 19th century.

- The decade of the 1990s was very likely the warmest decade in the instrumental record, which dates back to 1861.
- On average, between 1950 and 1993, nighttime daily minimum air temperatures over land increased by about 0.2°C per decade. This has lengthened the freeze-free season in many mid- and high latitude regions.
- It is very likely that precipitation has increased by 0.5 to 1.0% per decade in the 20th century over most mid- and high latitudes of the Northern Hemisphere continents, and it is likely that rainfall has increased by 0.2 to 0.3% per decade over the tropical land areas. It is also likely that rainfall has decreased over much of the Northern Hemisphere subtropical (10°N to 30°N) lands areas during the 20th century by about 0.3% per decade. No comparable changes have been observed over the Southern Hemisphere land areas.
- Global sea level has risen by between 4 and 8 inches (0.1 and 0.2 meters) over the past 100 years, and much of the increase is thought to be related to the rising global average temperature.

Much climate research has focused on determining whether the human-caused increase in greenhouse gases is the cause of the observed changes in the last century's climate. As noted in the most recent climate change report of the IPCC (2001):

There is new and stronger evidence that most of the warming observed over the last 50 years is attributable to human activities.

The 2001 NRC report noted that:

The changes observed over the last several decades are likely mostly due to human activities, but we cannot rule out that some significant part of these changes is also a reflection of natural variability.

These statements reflect the progress made in the recent decade in reducing the uncertainties associated with quantifying the factors that influence climate, along with the difficulties in distinguishing between human-induced climate change and natural variability. Underlying recent progress are: a longer and more closely scrutinized temperature record; improved model estimates of climate variability; reconstructions of climate data for the past 1,000 years; new estimates of the climate response to natural and human-caused forcing; improved understanding of the driving forces of the global climate, in both the tropics and the polar regions; and new techniques for the detection of change and the attribution of change to causal factors.

RESPONSES OF FORESTS TO VARIATION IN NUTRIENT AVAILABILITY

One of the more challenging problems for eco-physiologists is to assess the supply of nutrients available to sustain forest growth. A little insight is gained by simply analysing the elemental content in the soil because many nutrients are present in unavailable forms, and trees differ in their abilities to extract essential nutrients from throughout the soil profile. In addition, a range of nutrients can be acquired from the air via aerosol deposition on foliage and additional nitrogen can be acquired through root associations with symbiotic nitrogen-fixing bacteria. Physiologists and chemists, starting with Liebig in 1840, have established those elements essential for growth, defined specific deficiency symptoms and related concentrations in foliage to minimum requirements. We emphasize two additional components of nutrition that have significant impact on the growth of forests: balance and internal recycling. We discuss how nutrient imbalances increase the susceptibility of forests to insects and pathogens.

Optimum Nutrition

Torsten Ingestad (1979) developed a hydroponic system where the balance and total amount of nutrients could be varied daily to maintain relative growth rates constant over short periods. In a series of experiments he defined 'optimum nutritional balance' as that state when the ratios of nutrients in solution and leaf tissue maintain similar values. Under controlled laboratory conditions, the optimum nutrient balance, relative to the concentration of nitrogen, is similar for a wide range of tree species.

Further refinement in understanding and assessment of nutritional status is gained by defining 'metabolically active forms' in plant tissue. For phosphorus, the inorganic form best reflects its availability.

For nitrogen, most reserves are in the form of proteins. Above a certain concentration, excess nitrogen accumulates as free amino acids. This may be symptomatic of a generally unfavourable physiological status or it may be evidence that the rate of photosynthesis does not depend exclusively on substrate concentration.

Sune Linder, a compatriot of Ingestad, extended these optimum nutrient addition experiments into the field, providing convincing evidence that the growth of boreal forests of pine and spruce are more limited by slow release of nutrients through decomposition of organic matter than by climatic restrictions on photosynthesis and growth.

Internal Recycling of Nutrients

It has been long recognized that evergreen forests can accumulate large amounts of foliage in excess of that required for efficient interception of solar radiation. It is also known that a linear correlation exists between the total nitrogen content in foliage and wood production. By comparing the nutrient content in leaf litterfall with that in new and older foliage, it has been estimated that initially well-fertilized plantations can maintain favourable growth rates over decades by recycling stored nutrient capital acquired during development of a dense canopy. Ecophysiologists have contributed to our understanding of nutrient uptake and internal recyling by defining seasonal patterns more precisely and by identifying the dynamic exchange from one internal pool to another. Insights gained from such analyses include recognition that:

- fast growth of evergreens results in a greater proportion of shoot nitrogen reserves going into new growth at the expense of older foliage.
- when low growth rates prevail, elements in short supply are preferentially translocated to new growth.
- deciduous trees may obtain more than 50% of their nitrogen from internal sources.
- during periods of rapid leaf expansion, uptake of nutrients from the soil plays a minor role in meeting demands for deciduous as well as evergreen trees.
- understorey vegetation may contribute disproportionately to the acquisition and recycling of nutrients within forest ecosystems.

These insights have important implications in the management of vegetation and in the assessment of forest ecosystem response to atmospheric deposition of nitrogen and sulphur-rich compounds. For example, significant changes in the ratio of calcium to aluminium and the concentrations of essential trace elements in annual growth rings of trees bear witness to the impact of air pollution.

CLIMATE CHANGE AND THE GLOBAL CARBON CYCLE

In 2001 the IPCC concluded that most of the warming observed over the last half of the twentieth century can be attributed to human activities that have increased greenhouse gas concentrations in the atmosphere. They also warned that these changes will continue to drive rapid climate changes for several centuries to come. Chief amongst these greenhouse gases is CO_2, whose atmospheric concentrations have been dramatically altered by human perturbations to the global carbon cycle.

Over at least the 420 000 years prior to the twentieth century, the atmospheric concentration of CO_2 only varied between ~ 80 ppmv (parts per million volume) (during the glaciations when the global temperature was 8 - 9 °C colder than today) and ~ 180 ppmv (during the interglacial periods where the temperature was similar to present values). This range of variation in atmospheric CO_2 is remarkably narrow, given that its concentration is determined by a highly dynamic biogeochemical cycle.This suggests that the global carbon cycle was controlled by powerful biological feedback processes to maintain a close balance between net photosynthetic uptake of CO_2 by the biosphere and its total respiration-the net source and sink strengths of the biosphere was very close to zero over at least the last 420,000 years.

There is convincing evidence that the biosphere has played a major role in regulating Earth's climate. In particular, recent data shows that although there have been periods during which Earth's temperature changed abruptly without discernable accompanying changes in the atmospheric CO2 concentrations, the converse does not appear in the glacial-interglacial records. The warming from glacial to interglacial conditions was relatively rapid, while the cooling phase leading to glaciation was initially rapid (possibly suggesting perturbation by an external event), but eventually gradual (indicating strong feedbacks that act to counter the change). These patterns suggest a long-term asymmetry in the global rates of CO_2 uptake and release by the biosphere. The terrestrial and ocean ecosystems act as buffers to maintain the global temperature in a habitable range. In contrast to the long-term record, the atmospheric CO_2 concentration today is ~ 370 ppmv - nearly 35% higher than at any time in the past 420 000 years - as a result of human perturbations to the global carbon cycle. The concentration is also rising at a rate that is at least ten times and perhaps as much as one hundred times faster than ever before observed. Clearly the biosphere's regulation of the global carbon cycle - and hence the climate system - has changed. Although it is straightforward to quantify the direct anthropogenic inputs of CO_2 to the atmosphere, a quantitative explanation of the rates of atmospheric increase has proven immensely challenging, precisely because of the strong feedbacks exerted by terrestrial and ocean ecosystems to the changes. Understanding of the biospheric feedback - the response of the world's biota to the perturbations - is needed both to gauge the magnitude of future impacts and to design appropriate mitigation actions.

HUMAN PERTURBATIONS TO THE GLOBAL CARBON CYCLE

Human perturbations to the carbon cycle have been both direct and indirect. Obvious direct effects are the addition of new carbon to the active[2] global carbon cycle through the combustion of fossil fuels, and the modification of the vegetation structure and distribution through land-use change. Deforestation, the removal of forest vegetation and replacement by other surface cover, has the largest land-use change impact on the carbon cycle, both through the loss of photosynthetic capacity in forest vegetation and the simultaneous release of large carbon stocks accumulated in forest ecosystems over long periods of time. Indirect human impacts on the carbon cycle include changes in other major global biogeochemical cycles, alteration of the atmospheric composition through the additions of pollutants as well as CO_2, and changes in the biodiversity of landscapes and species.

Currently about three-quarters of the direct human perturbations to the global carbon cycle are due to fossil-fuel combustion, emissions of which currently exceed 6 Gt C/yr (gigatonnes of carbon per year) and are still increasing. To provide perspective, this emission is equivalent to the total incineration of half of all the trees in Canada - with no residues, charcoal or soot. Every year. Since the mid-nineteenth century, however, the cumulative addition of anthropogenic CO_2 to the atmosphere by land-use change has been nearly as large (~ 156 Gt C) as that from fossil fuel use (~ 280 Gt C), and continues to be an important anthropogenic emission (2.2 Gt C/yr).

Of the 7.6 ± 0.8 Gt C/yr of CO_2 added to the atmosphere by human activities during the period 1980 to 1995, only 3.2 ± 1.0 Gt C/yr remains there, with the rest taken up about equally by the oceans and by terrestrial ecosystems. Earth's biosphere thus actively removes some of the new carbon that humans have added. Terrestrial ecosystems, in particular, sequestered (took up and retained) 2.3 ± 0.9 Gt C/yr, even after accounting for the loss of 2.0 - 2.2 Gt C/yr from deforestation.

Of pressing importance in the search for mitigation strategies - activities that slow or reverse the buildup of atmospheric CO_2 - is the understanding of the mechanisms responsible for the present net biospheric uptake. Will these mechanisms continue to offset direct anthropogenic emissions? Or will they decline in strength, or even fail entirely as the carbon cycle-climate system moves into a new mode of operation, as several new terrestrial and ocean model simulations alarmingly suggest?

THE ANATOMY OF SOURCES AND SINKS AT THE STAND AND LANDSCAPE SCALE

A forest ecosystem acts as a "sink" (a net removal of atmospheric CO_2) when there is an increase in the sum of all carbon stocks retained in the forest vegetation itself and the derived stocks of organic carbon in other reservoirs. The most important of these derived reservoirs are the detritus and soil organic matter pools of the forest ecosystem.

Figure 3 shows the conceptual pools and transfers of carbon involved in forest ecosystems and in the forest sector. In addition to the ecosystem compartments (vegetation, detritus and soil pools) and the exported pools that are located off-site (including forest products and the waste created during their manufacture and abandonment in landfills), the use of fossil-fuel reserves by the forest sector is also shown.

The net carbon balance in a forest ecosystem (NEP) can be estimated by summing all the changes in ecosystem carbon stocks (the 'inventory' method), or by direct measurement of the net exchange of CO_2 with the atmosphere (e.g. eddy correlation techniques). Provided all stocks and fluxes are accounted for, the two approaches must give identical answers (conservation of mass). In practice, a combination of the two approaches is used.

The net carbon balance of a given stand of trees (patch) varies with the prevailing conditions (affecting rates of CO_2 uptake and release) but also depends strongly on the stage of development of the stand and its past history. At the landscape (or biome) scale, a forest is a mosaic of many stands of trees (individual ecosystems) in various stages of development. The net carbon balance at this scale is the summation across all such ecosystems in the landscape.

Change in the net carbon accumulation at the landscape scale thus has two components: changes in the productivity of the individual ecosystems with environmental variations and the changes in the age-class distribution associated with landscape variation in mortality and recruitment. The age-class distribution is a histogram of the fraction of the forest in each age class (typically 10 or 20 years), and is a record of past mortality and recruitment events.

Landscape sources and sinks of atmospheric carbon can arise from changes in ecosystem productivity or from changes in the disturbance regime. If the disturbance rates increase, the age-class distribution shifts to the left (younger stands) and the total carbon retained in the ecosystems in the landscape decreases - the landscape is a transient net source of CO_2 to the atmosphere until a new stable age-class distribution is reached. (If carbon is transported out of the ecosystem landscape to decompose in off-site reservoirs, such as forest products, the landscape source is reduced by that amount - in essence this component of the source is exported.) Similarly, if disturbances are suppressed, the distributions shift to the right (older forest stands) and carbon stocks increase with a transient net removal of CO_2 from the atmosphere.

3

Land Uses to Climate Variability and Climate Change

INTRODUCTION

Few areas in the world are more vulnerable to climatic variability than the small islands and low-lying states in the Caribbean Basin. While they have a natural vulnerability to hurricanes and tropical storms because of their small land masses, that vulnerability has been exacerbated because of human activities — these include intensive land development and high population density in coastal zones, poorly developed coastal infrastructure, limited human and cash reserves, impacts of tourism-based industries, and a lack of trained personnel to address the impacts of climate variability.

Environmental degradation in many of these Caribbean islands and low-lying states has increased the risk that climate change is predicted to have on the already-limited natural resources and economies. According to a 1998 Inter-American Development Bank study, land use and resource allocation conflicts in the coastal zone, degradation of coastal ecosystems, fish stock reduction and declining water quality from land-based activities have been intensifying the demands on natural resources. Such demands limit the ability of resources such as coral reefs and aquifers as well as the populations and economies that depend on these resources to adapt to climatic changes. Thus, the current ecological challenges faced by these low-lying and island states are also increasing their vulnerability to climate change, and the potential damages associated with climatic impacts.

The climatic impacts likely to be the most significant and immediate are: (1) changes in varying rainfall regimes and soil moisture budgets, (2) potential increases in the frequency and intensity of hurricanes, and (3) changes in regional and local sea levels and patterns of wave action. Rising sea levels, in particular, pose threats of severe ecological disruption to Caribbean social structures and economies, which are highly concentrated in the coastal zone (*i.e.*, within two kilometers of the coast). Most islands have been caught in a Catch-22: on the one hand, their economies have become largely dependent on the tourism industry, which has featured pristine beaches, coral reef ecosystems, and other coastal amenities; and yet to attract and continue to attract international tourism, many island states have compromised their natural assets in an effort to remain competitive. In the process, other economic activities such as agriculture and manufacturing industries have become jeopardized, in part because of the strain in a number of island states on resources such as potable water, coral reef habitats, and fisheries.

To develop realistic strategies for trying to curb the impacts that human activities exacerbate in sea level rise, the Coastal Planning for Adaptation to Global Climate Change (CPACC) project first identified distinct physical impacts that are expected to occur with climate change. They include:

- submergence of low-lying wetland and dry land areas;
- erosion of soft shores by increasing offshore loss of sediment;
- increased salinity of estuaries and aquifers;
- rising coastal water tables; and
- increased and more severe coastal flooding and storm damage.

Because more than 40 percent of the population lives within two kilometers of the coast, these impacts will affect many aspects of Caribbean coastal economic development. Strategic efforts to reduce this vulnerability are consistent with meeting the ecological challenges that small island coastal ecosystems are already having to contend with. These ecosystems and fringe landforms — among them, mangroves, wetlands, coastal flood plains, forests, dunes, berms —) serve as natural shock absorbers for protecting coastal infrastructure and land uses, such as agriculture, against tropical storms and hurricanes; they also provide critical storage capacities for storm surges and floodwaters. When the functioning of these coastal and fringe systems is compromised, the coastal zone is even more vulnerable than it ordinarily would be. There is increasing evidence that strong measures to conserve and enhance their ecological resilience to the stress of sea level rise, for example, through effective land use planning (*e.g.* set back requirements, or transferred development rights) can help reduce the vulnerability of these small island states to looming disaster. Doing so will require concentrated efforts by island governments to develop coordinated policies that have the goal of managing human impacts on climate change impacts, particularly sea level rise.

The need to act decisively in protecting coastal zones must contend with the demands of the immense tourism industry, which is likely to remain central to Caribbean island economies. It must be recognized, however, that if environmental degradation becomes so great, the islands may no longer be as attractive, and therefore competitive, with other tourism destinations. Key challenges facing the Anglophone Caribbean states are how to balance economic need that the tourism industry fills with minimizing the environmental stress that it can bring. To do so means dealing innovatively both with current land development — in order to reverse as much as possible the damages it has already caused — and projected land development, so that environmental impacts can be minimized. Meeting such challenges may be difficult, in part because economic investment generally does not project over the long term that environmental policy must be aimed at. However, there is great potential for reversing current environmental impacts and those projected for the future — it will take political will and cooperation among the Caribbean island states to undertake actions that can help achieve these aims.

This chapter sets out a realistic approach for developing strategic approaches to meeting the challenges of sea level rise by minimizing the impacts of human activities. A background on the physical and economic characteristics of the Caribbean basin. The human impacts that are increasing the vulnerability of island ecosystems. A series of recommendations that Caribbean island governments can employ to meet the challenges ahead.

BACKGROUND: PHYSIOLOGY AND ECONOMIC DEVELOPMENT

GEOGRAPHIC DESCRIPTION

The Wider Caribbean Region comprises an area including the marine environment of the Gulf of Mexico, the Caribbean Sea and adjacent areas of the Atlantic Ocean and the surrounding coastal regions. The two main bodies of water of the region - the Gulf of Mexico and the Caribbean Sea -form a partially enclosed sea with a total surface area of about 4.24 million square kilometers. This region has an ocean circulation pattern that follows an east-to-west movement in the Caribbean Sea, followed by a southeast-to-northwest movement into the Yucatan Basin and, finally reaching the Gulf of Mexico, eastward again through the Florida Straits. The region is fed by the Northern Equatorial Current, which flows northward looping into the Gulf of Mexico. Eighteen important rivers from nine countries are part of the region's basin, including large tributaries such as the Mississippi River. The continental shelf in the region is mostly narrow except in few areas, such as off the coast of Belize, Nicaragua and the Bahamas. The surface water temperature average of some 27°C generally varies ±3 degrees. The stability of warm water conditions has fostered the region's unique and highly productive ecosystems of coral reefs, seagrass beds and mangroves, making the areas close to the shores highly productive. About 14 percent of the world's coral reefs are found in the Wider Caribbean, with minor areas around the islands and major barrier reefs in the Yucatan Peninsula and offshore Belize, Honduras and Nicaragua.

The countries considered in this chapter. The Bahamas, for example, has the greatest coastline by far, while Guyana is the smallest, relative to its land mass (0.005 km/km2). Jamaica and St. Kitts and Nevis are the most and the least populated countries, respectively; however, in relation to population density, Barbados is the most densely populated country at 640, while Guyana and Belize are the least densely populated, at 3.2 and 11.

MAJOR ECONOMIC ACTIVITIES

The Anglophone Caribbean region and their influence influence on natural resources. Tourism is presently the major economic activity in the region — an estimated 12 million tourists visit the region each year and another eight million visit in cruise vessels , which is subject to global economic fluctuation and competition. Tourism accounts for between 25-35 percent of the total Caribbean region economy, it represents one-quarter of the foreign exchange earnings, and it provides about one-fifth of the jobs. The Small Island Developing States - Progrmme of Action identifies the tourism industry as a priority for development, and as one of the main pressures on the environment.

The small size and rugged topography of most countries in the Anglophone Caribbean limit the different types of land use. The needs of tourism compete with needs of housing, agriculture, industry, roads, and ports. According to FAO (1999), among the countries considered in this project, Jamaica is the country with the most agricultural land use, and Dominica the one with the most forest and wood land use. Other countries such as the Bahamas, Antigua and Barbuda, Grenada and St. Lucia have a large percentage of their land dedicated to other uses including buildings and urban facilities. Land-use management has been identified as the key issue for forestry development in the Anglophone Caribbean. The area of native forest in the region has declined continuously, and the average deforestation rates vary between 0.8-7.2% per year, although the rate of deforestation has slowed in recent years. It is interesting to note that islands such as Barbados and Antigua and Barbuda have lost all their native forest to urban and coastal development. This phenomenon has had a significant impact on polluted runoff to the near-shore environment.

While freshwater resources in the Anglophone Caribbean region are generally scarce compared with other island regions in the world , they differ depending on the geology, topography and size of the island. Many small islands in the region have virtually no fresh water ecosystems. Barbados, for instance, is considered one of the top ten most arid countries in the world, according to the World Health Organization, while St. Kitts has large freshwater reserves. Expansion of the tourism industry and agricultural needs have greatly increased the demand of freshwater; however, it is difficult, except in the cases of big islands, to evaluate the magnitude of water reserves for most of the Caribbean Islands.

Marine and coastal regions harbor the most important natural resources of the Caribbean region. They not only sustain some of the most diverse ecosystems in the world (*e.g.* coral reefs, seagrass beds and mangroves), but they also are linked to most of the activities in the region from food generation to tourism demand. Considering the Caribbean islands' exclusive economic zone areas versus their land areas, most of these islands have very small water-land ratios, with an average for all the Caribbean of about 1:30. This ratio illustrates the importance of the marine environment in the region, and the critical need to incorporate these environments into development planning agendas. Similarly, because of the uniqueness of the Caribbean Sea circulation patterns, each island's impact on the near-shore environment is directly related to the level of anthropogenic activity landward and can have a cumulative effect on the entire region.

LAND-USE PLANNING IN CARIBBEAN STATES

Antigua and Barbuda.– Land-use planning and development controls in Antigua and Barbuda generally are weak and underdeveloped, largely because of underlying political suspicion of the land-use planning process and limited technical expertise.

The consequence has been coastal urban sprawl, soil erosion, inadequate land and solid waste disposal in certain areas, steady conversion of agricultural land to residential and other purposes, and costly provision of public infrastructure and utilities. These trends have increased the vulnerability of Antigua and Barbuda to flooding from extreme weather events such as hurricanes and the projected inundation from rising sea levels. In order to reduce this vulnerability, coordination in land-use planning will be essential to the overall process of development.

St. Kitts and Nevis. While the coastal rural areas of St. Kitts are characterized by ribbon developments, none of which can be said to have occurred as part of a planned physical development strategy, there are a handful of locations where planned developments have been implemented. The drift from rural to urban areas over the past two decades has led to Bassaterre becoming overcrowded. Sandy Point and Canyon are also experiencing commercial and industrial expansion on the margins of the town. For the most part, the traditional suburbs are poorly planned and present formidable challenges in terms of adequate housing, services and infrastructure provision.

Thus, inadequate coastal development is a big problem in St. Kitts and Nevis. In addition, the types of coastal development and infrastructure on the islands are particularly vulnerable to extreme weather events and climate change: most public buildings have been built to withstand hurricane force winds of 160 kph (100 mph), and in some cases up to 240 kph (150 mph). However, it is unclear to what extent such facilities will be able to meaningfully stand up to the type of intensified hurricane force winds projected as a result of climate change, and perhaps foreshadowed in hurricanes such as Mitch in 1998 and Michelle in 2001. These experiences have indicated, that tourism structures and facilities are particularly vulnerable to damage associated with extreme events (St. Kitts-Nevis Government, 2001).

St. Lucia. For many years, urban growth in St. Lucia occurred in the absence of a proper planning regime. Moreover, even with specific efforts to address this situation beginning in the early 1970s, the pre-existing conditions, the inadequacy of planning legislation itself, and the difficulties of enforcing this legislation have all led to intensive and relatively uncontrolled development in coastal areas.

LOCATION OF COASTAL INFRASTRUCTURE IN HAZARDOUS AREAS

Coastal urbanization and land-use controls were not seen as critical issues in the Caribbean so that lands have been inefficiently utilized on the free-hold and leasehold land use allocation system. This has led to many sensitive areas being developed, particularly those lands with high environmental values such as offering watershed protection to possessing high biodiversity. Most of the coastal open space that could easily be developed has long since been built upon, and the continued growth of the population has meant that hazard-prone areas such as coastal flood plains or mountain slopes are increasingly being used for settlements. Currently, a majority of the population centers, agricultural areas, tourism, ports and other major sites for industrial and commercial activity are located in the coastal zone, many in low-lying floodplains that are vulnerable to flooding from hurricanes and sea level rise. Even islands where changes in elevation are significant have a substantial share of their population and capital investment in vulnerable nearshore areas. These development patterns and pressures can be expected to intensify with the growth of coastal urban areas, as increasing demand for housing and infrastructure result in greater demand for areas in hazardous zones. This pattern might increase the risk in hazardous areas, due to the removal of natural protection (*e.g.* magroves) against extreme weather events.

In many areas in the region, the human settlements and infrastructure located in hazardous nearshore areas are the types of structures likely to be most vulnerable to climate variability and change. Informal houses and squatter settlements are increasingly being built in coastal lowlands or on hill slopes. This is in part a result of landholding patterns referred to previously: wealthy landowners have always controlled vast plantations and large tracts of land, while small farmers typically own one hectare of marginal land, often on steep slopes. Consequently, those types of structures and settlements, *i.e.,* small farms, informal housing and squatter settlements, are often relegated to the most vulnerable or hazard-prone areas, and are the hardest hit by extreme weather events. These trends towards development in vulnerable nearshore areas in the Caribbean are a cause for concern, because such developments carry a high risk that can only be partially ameliorated by structural measures.

In general, coastal infrastructure in the region is likely to be most vulnerable to extreme weather events such as hurricanes, which may increase in intensity as the climate changes, and sea levels rise. Additionally, such developments also have a large impact on coastal habitats, as evidenced by wastewater discharge, increased erosion, siltation of coastal waters and clearance of mangroves and littoral forests, all of which have been associated with urban areas and infrastructure.

Coastal Infrastructure in Hazardous Areas

Antigua and Barbuda. The importance of the location of coastal infrastructure was evident during the 1995 and 1999 hurricane seasons, when the country was hit by two hurricanes in the space of one month. Heavy rains and storm surges produced localized flooding and landslides. A great deal of damage inflicted on the houses and public buildings by Hurricanes Luis and Lenny can be traced to inappropriate standards for methods and materials used in construction, and inadequate land-use planning that allowed development in hazardous locations along the shoreline. Particularly troubling is the number of wells and sinkholes that are often located along the coast in Antigua and Barbuda, and in many countries throughout the Caribbean. For many areas, the depth of the water table is generally less than a few meters in the coastal lowlands, and will likely be affected by saline intrusion under projected levels of sea level rise. To further illustrate the vulnerability of coastal infrastructure and settlements in Antigua and Barbuda, a recent assessment of the largest human settlement revealed that 19% of the settlement was classified as high vulnerability, and another 19% was evaluated as very high vulnerability (Antigua and Barbuda Government, 2001).

Barbados. All of the principal urban settlement (including that of the capital, Bridgetown) is located within the coastal urban corridor. Approximately 25% of the island's population lives within two miles of the coast and is concentrated within a continuous linear urban corridor, extending for the entire length of the west and south coasts. Since the 1990s, there has been increasing demand for residential housing development, and the coastal urban corridor has expanded significantly along the south coast, with a change of land use from agricultural lands to subdivisions for residential purposes. Many homes are located in floodplains and low-lying coastal lands. Around the capital of Bridgetown, for example, there exists much low-income 'slum housing', which sits on what is in essence low gradient coastal land that forms part of the Constitution River catchment. There are several water supply wells and mains along the south and west coasts that are vulnerable to saline intrusion, and the two main coastal highways running out of Bridgetown are both barely above sea level (Barbados Government, 2001).

Belize. – Forty-five percent of the population currently resides in the coastal zone, and in Belize City residential areas such as Vista del Mar, Bella Vista,

Belama and Fort George have been constructed on drained and reclaimed coastal wetlands which are vulnerable to flooding resulting from hurricanes and projected sea level rise. Similarly, infrastructure in most of the other communities like Dangriga, Corozal, the Placentia Peninsula and Ambergis Caye are currently threatened even by a 20 cm sea level rise. Throughout the four coastal districts of Belize, large urban and rural infrastructure is prone to seasonal flooding and inundation, as all of these districts have human settlements concentrated in areas that range from 0 to 5 meters above the present sea level.

St. Lucia. Characterized by a high concentration of hotels, ports, roads, settlement along the coast, often in low-lying reclaimed areasa majority of St. Lucia's population and critical infrastructure is located on low-lying reclaimed coastal land. This situation has made the country extremely vulnerable to extreme weather events and projected sea level rise. For example, over 60% of the entire population resides along the Castries-Gros Islet corridor on the northwest of the island, where much of the settlements are built upon low-lying coastal lands prone to flooding. St. Lucia's two airports and many roads lie on flat coastal lands, and in some cases, a few meters from the sea. Similarly, some squatter settlements are erected in floodplains or in areas cleared of mangroves.

INADEQUATE WASTE DISPOSAL SYSTEMS

The management of wastes throughout the Caribbean is marginally effective. As a result, severe damages to coastal ecosystems that should function as buffers against the impacts of climate variability and change have rendered coastal infrastructure and land uses increasingly vulnerable to climate change and sea level rise. Inadequate institutional arrangements, ineffective legislation, lack of monitoring of effluents and emissions, and failure of enforcement mechanisms relating to the management of waste characterize almost all countries in the Caribbean.

Solid Waste Disposal – Many of the landfills in the Caribbean are sited in hazard-prone areas, without any of the necessary hydro-geological investigations. Drains, rivers, streams, wetlands, alleys and land depressions are all used as waste depositories. Many informal dump sites are unauthorized and were created in response to excess demand for waste disposal facilities. As a result, coastal ecosystems have often been contaminated, particularly mangroves and wetlands. In St. Vincent and the Grenadines, for example, lack of siting criteria for dumps has led to floodwaters taking sediments and plastics out to sea and depositing them on reefs (St. Vincent and Grenadines Government, 2000). Tourism has also increased the amount of solid waste in the region, due to the massive import of goods to service the industry that generates additional demand for informal and unauthorized waste disposal facilities.

Liquid Waste Disposal – Poor water quality is one of the biggest threats to the reefs that buffer flooding from hurricanes and rising seas. Trends in the Caribbean include increased generation of wastewater, mostly on-site septic tank, soak-away and package treatment systems. Standards for liquid waste disposal are minimal and poorly monitored and enforced. Treatment plants have a poor operational status – 75% do not meet minimum standards, only 16% have a certified operator, less than half have monitoring data available, and 33% of treated wastewater from hotels is discharged to the marine environment; these data are based on a 1991 study of 13 CARICOM countries (Caribbean Environmental Health Institute, 1992). In some cases, the tourist sector shares public sewage infrastructure with local communities, which increases the discharge of untreated effluents into the coastal environment. According to the Caribbean Environmental Health Institute, the Caribbean average water consumption in the hotel sector is 3 to 4 times higher than that of domestic consumers, and according to the Caribbean Tourism Organization , 80 to 90% of the sewage generated across the region by hotels is disposed of in nearshore coastal waters without adequate treatment. The end result of primary treated and untreated sewage disposal at the nearshore is the continuing build-up of algae on reefs. Similarly, such poor sewage disposal has also resulted in growing levels of eutrophication in many coastal waters near urban areas and harbors throughout the region. Traditionally, emphasis has been on low cost treatment approaches, such as septic systems or sewage outfalls. In the majority of these islands, the tax base has been insufficient to cover the expenditure necessary to establish treatment systems or to develop the infrastructure to facilitate effective sewage septic connections.

Construction and Insurance Incentives

The Caribbean property insurance industry is characterized by a proliferation of general agency units representing foreign companies sharing the market with a relatively small number of Caribbean-owned companies. The number of agencies is disproportionately large for the small volume of property risk underwriting that actually occurs in the region. In addition, the proportion of the risk of extreme weather events that is retained by the companies in the region is small, estimated at 15%, with the remainder being ceded to reinsurers outside the region. As a consequence, competition for agency fees and reinsurance commissions tends to drive the underwriting practice, at the expense of sound appreciation for the risk of buildings and structures, to extreme weather events and future climatic changes.

Because few insurers provide incentives for property owners and builders to adopt measures that would mitigate the effects of extreme weather events and sea level rise, and few property owners do so voluntarily, the quality of buildings and structures throughout the Caribbean is not adequate. Currently, cost competition prevents developers, designers, and contractors from building safer and less vulnerable structures, in the absence of sound financial incentives and stronger building codes and standards. This trend towards building vulnerable structures is exacerbated by a low public perception of risk and vulnerability to climate variability and change in many countries.

The bottom line is that property insurance in the Caribbean has several different faces, and its most common face is not pretty. The standard product offered by the insurance industry to the average property owner is expensive; more than half of the premium paid by the insured is allotted to commissions, profit, marketing, and administrative expenses. The underwriter pays little attention to catastrophe risk and vulnerability to climate variability and change, and the industry does not offer the insured any incentive to reduce that vulnerability. A substantial part of society is uninsured (*i.e.*, squatter settlements and informal housing), and this applies not only to the lower income sectors but also to a large majority of government-owned properties.

Destruction of Ecological Buffer Systems

A diverse array of stresses resulting from coastal infrastructure and land-use choices are increasing the vulnerability of natural systems in the Caribbean and reducing their effectiveness as buffers against extreme weather events and sea level rise. Deteriorating water quality of coastal zones is threatening the health of both reef and mangrove systems and rendering them more vulnerable to extreme events. For example, shoreline construction of harbors, dikes, channels and other related activities have all resulted in excessive sediment loads to the coastal waters, reducing the ability of seagrasses and coral reefs to get the light they need for photosynthesis.

Mangroves have been cleared in large measure throughout the Caribbean to accommodate coastal development. For example, most mangroves in St. Lucia are not protected, and many have been lost along the west coast to tourist developments. Similarly, in Belize mangroves are being lost to urban expansion for residential development and coastal subdivisions. The loss of mangroves deprives coastal infrastructure and land uses in the Caribbean of much-needed ecological buffers to flooding. Furthermore, where mangroves have been cleared the oxidation of peat can lead to a decline in land level, making such areas even more susceptible to flooding. Similar to mangrove areas, many wetlands throughout the Caribbean have been filled and destroyed for the construction of tourism and residential developments.

Beaches throughout the Caribbean — among the great attractions of the tourism industry — are increasingly at risk from sand mining activities, which exacerbate erosion of these landforms and increase the vulnerability of coastal infrastructure to flooding. Traditionally in the region, sand used to fill coastal wetlands and lowlands has come from beaches, but now the current demand in many countries far exceeds the capacity of the beaches to provide it, particularly without increasing erosion. This trend should be of concern to land use planners, because in order for coastal infrastructure in many communities and countries to withstand rising sea levels, a constant supply of large volumes of sand and sediment will be required. However, sand mining for coastal construction is significantly reducing one of the major sources of this sediment, and rapidly increasing the vulnerability of coastal infrastructure and land uses to flooding from hurricanes and sea level rise.

Land-Use Planning

Land-use planning in the Caribbean is often sector-oriented, with little capacity to respond to local level needs and conditions. These top-down institutional approaches to land-use planning have largely been inherited from previous colonial experiences, and are ill equipped to accommodate community responses to climate variability and change. In many cases, development choices are made with little or no consultation with affected stakeholders, and often result in settlements and structures in vulnerable or hazardous areas. Throughout the Caribbean there is a need to strengthen the links between local communities and national governments so that the communities increasingly gain a voice in planning and land-use decisions.

DESTRUCTIVE AGRICULTURE AND FORESTRY PRACTICES

The coastal zone is not an isolated system, and is impacted by other land uses. Rivers and waterways carry pollutants and sediments resulting from inland activities to the coast. Similarly, agricultural and forestry practices are recognized sources of sediments and chemical pollutants to the marine environments in the Caribbean. For example, pesticides and fertilizers used on farms often reach adjacent coastal waters and reservoirs via rivers and other waterways, impacting coral reefs, seagrasses, and fish habitats. Similarly, agricultural practices such as extensive land clearing, the removal of topsoil resources, and replacement of natural vegetation or forest exacerbates droughts by allowing the soil to dry out more quickly. These droughts increase the risk that fires, which are set deliberately to clear land for agricultural purposes, can spread out of control. This practice decreases the fertility of the soil and increases the land erosion due to rain runoff.

IMPACTS OF CLIMATE VARIABILITY AND CHANGE ON COASTAL INFRASTRUCTURE AND LAND USES

Given all of the trends identified previously, it is possible and indeed likely, that coastal infrastructure and land uses in many of the countries in the Caribbean will be significantly impacted as a result of climate change. According to the IPCC (2001), the impacts resulting from climate change are expected to include the following:

- Average temperature rise in the region of approximately 3ºC by 2080
- Average sea level rise of 5 mm/year over the next century
- Increased intensity and potentially increased frequency of extreme weather events such as hurricanes
- Changing patterns of rainfall.

The effects of these climatic impacts upon coastal infrastructure and land uses in the Caribbean will likely be severe. Most capital cities and major settlements have been developed in low-lying coastal areas that are vulnerable to flooding from hurricanes and sea level rise.

This coastal infrastructure will only become more vulnerable to flooding as the climate changes and sea levels rise.

Rising sea levels will provide a higher base for storm surges to propagate further inland, enabling weaker storms to have an effect on infrastructure that would otherwise only be affected by stronger storms. Some of the specific effects of climate change on coastal infrastructure and settlements will likely include:

Coastal Infrastructure and Local Economy

- Destruction and loss of infrastructure in low-lying coastal areas arising from sea level rise;
- Greater property damage and loss from increased hurricane activity;
- Flooding of low-lying areas from storm run-off and storm surge;
- Increased costs of sea defense mechanisms;
- Increased costs and reduced availability of insurance coverage for property;
- Dislocation of coastal populations, particularly squatter communities;
- Destruction of hotels and other beach front properties due to the effects of sea level rise, storm surge and hurricanes;
- Increased demand and competition for coastal lands as a result of land lost to sea level rise;
- Reduced demand for tourism and subsequent changes in islands' employment structure;
- Changing patterns of port development and infrastructure as a result of sea level rise that include higher maintenance costs and increased dredging;

Natural Resources and Habitats:

- Saltwater intrusion into coastal wells and other freshwater resources;
- Destruction of coral reefs as a result of bleaching from higher sea surface temperatures;
- Loss of mangroves to sea level rise in many areas where coastal topography, mangrove systems, and coastal infrastructure do not allow sedimentation to keep pace with rising sea levels, and,
- Increased coastal erosion.

Barbados, Potential Impacts of Sea Level Rise on Near Shore Infrastructure

Some individual examples and projections of these impacts include St. Kitts and Nevis, where an intensification of present patterns of coastal erosion, saline intrusion and saltwater inundation are expected as a result of climate change. For example, the storm surge associated with Hurricane Georges in 1998 had a tremendous adverse impact on tourism infrastructure in St. Kitts (cruise ship facilities) and in Nevis (hotel properties) (St. Kitts-Nevis Government, 2001). Similarly, in Barbados much of the coastal infrastructure is located in vulnerable areas likely to be inundated under projected rising sea levels (Barbados Government, 2000). This infrastructure includes:

- Barbados Light & Power Company Ltd.
- Barbados Port Authority
- Fort Willoughby (Coast Guard Station)
- Holetown Police Station
- Weston Fire Station
- Oistins Govt. Complex
- Government Headquarters (Ministry of Finance)
- Coastal Zone Management Unit
- Bayview Hospital
- Geriatric Hospital
- Port St. Charles Marina
- Desalination plant at Spring Garden
- Shell, Esso & Texaco Fuel Storage and Processing Facility at Spring Garden
- Major Coastal Highways

TRENDS INCREASING THE VULNERABILITY OF CLIMATE CHANGE

The vulnerability of coastal infrastructure and land-uses in the Caribbean to climate variability and change. Based on an understanding of this vulnerability, it will be possible to describe current and future climatic impacts to these sectors to identify adaptation responses to them. Because vulnerability results from both natural and anthropogenic processes, this section will examine trends that are increasing the stress of climate variability and change on coastal infrastructure and land uses. There are numbers of important differences among countries, in physical characteristics and population density. For these reasons, it is important to note that the general trends of increasing vulnerability described here should be placed in a country context; where possible we will give specific examples to illustrate these trends.

Throughout the region, several over-arching forces have shaped specific trends in vulnerability, such as increasing population densities and growth rates, the fast expansion of the tourism industry, particularly at the expense of agriculture, and uncontrolled coastal development and urbanization. These trends and others, coupled with inadequate policies and enforcement of those in effect, have at times resulted in the destruction of critical ecological buffers and increased the vulnerability of coastal infrastructure and land uses in the Caribbean.

INCREASING POPULATION DENSITIES AND GROWTH RATES

According to a recent FAO statistics (FAOSTAT) report, the overall population of the Wider Caribbean region is approximately 40 million. The population of the Caribbean almost doubled between 1950 and 1995, despite relatively high levels of emigration between 1960 and 1980. Over the same period, the population below age 15 declined, while the 60 and over age group increased. This trend is expected to remain the same to the year 2005. Factors contributing to this trend include the marked increase in life expectancy at birth observed over the past 35 years, with women living an average 4 to 6 years longer than men. As a result of this increase in population growth rates, greater coastal urbanization and population densities are occurring throughout the region, resulting in over-crowding in the coastal zones of many Caribbean countries, particularly with the movements from the agricultural to coastal tourism sectors.

GROWTH OF THE TOURISM INDUSTRY

The countries in the Caribbean share a common history of European colonization, slavery and plantation economies. Caribbean islands had the ability, under natural conditions, to support large-scale plantations, and thus produce products for world export. Lands that did not possess the conditions to support these plantations were transformed into small farms to produce crops for local consumption. As a result, monocrop agriculture developed as the major economic activity and land use from mid-19th to late 20th in the Caribbean islands.

However, over the years competitive disadvantages with other regional markets throughout the world have led to a gradual decline in some agricultural activities in the Caribbean. The islands have always attracted tourism; however, with the post-World War II growth of European and North American economies and more disposal income by a greater number of people, the Caribbean became more attractive to more people. Tourism is now the region's largest growth industry, particularly coastal tourism (Caribbean Environmental Network, 1997). In addition, the geographic proximity of the Caribbean to North America and Europe has fostered a large inflow of foreign private hotel investment and aid-financed air, sea and inland transport infrastructure – the capital base of the tourism industry. The result has been a shift in the economic base of most countries from agriculture to tourism and associated services, which account for 25 to 35% of the total economy of the region, and 20% of all jobs (Caribbean Environmental Network, 1997). This change has also increased the average per capita income and wealth of the region. Between 1975 and 1995 the average per capita income in the Caribbean grew more than 30%, and in those countries in which tourism and financial services are well developed the per capita incomes reached the highest levels.

This increasing demand for coastal tourism in the region has changed the demand for land-use and increased the price of coastal lands. Trends in coastal lands have been skewed towards high-density, mass-market tourism sites close to the water's edge, which have increased the competitiveness of tourism with other activities for coastal land. The result has been a ribbon growth pattern of development along the coast, mostly due to space limitations for coastal tourism and the associated infrastructure requirements such as transport links. Additionally, the growing concentration of tourist activities has increased the dependency of the Caribbean islands on tourism earnings, giving this activity priority in the use of coastal and marine resources and public infrastructure (Caribbean Environmental Network, 1997).

Unfortunately, the growth of tourism throughout the Caribbean in the absence of coordinated land-use planning and development controls has led to patterns of urbanization which are rendering coastal infrastructure increasingly vulnerable to climate variability and change. Because most of the region's tourism activities are concentrated in the coastal zone, much of the infrastructure has been constructed in coastal areas vulnerable to extreme weather events such as hurricanes and rising seas that are predicted to result from climate change.

How did this progressive degradation of coastal regions in the Caribbean occur? It is a tale of commercial demand for nearshore land and the expediency of short-term gains in jobs and economic growth. As tourism growth fueled intensive and uncontrolled development throughout the region, coastal infrastructure such as roads, seaports and airports by necessity had to expand into more vulnerable areas (Antigua and Barbuda Government, 2001). This spatial distribution of coastal settlements and infrastructure has for the most part, evolved through natural trends, with little or no land-use controls. What land-use planning exists has relied upon development control mechanisms that for the most part have not been enforced, and have yet to successfully include building codes or standards that incorporate protections against climatic impacts. Economic instruments and incentives, which are generally far more effective for controlling land use, have not been widely applied.

Thus, current land-use planning and design guidelines in many areas of the Caribbean provide no incentives for developers and their architects to recognize or accommodate vulnerability to climate variability and change. With effective controls lacking, it should come as no surprise that the quality of coastal infrastructure construction is unsatisfactory in many areas.

Throughout the region there has been a trend of shifting away from traditional architectural practices to more vulnerable forms of construction, such as the squatter settlements that proliferate in many urban areas. These settlements are generally characterized by substandard housing and inadequate infrastructure, and are particularly vulnerable to extreme weather events and sea level rise, as they are the least equipped to withstand high wind speeds and flooding. As coastal urbanization and uncontrolled development continue to increase in the Caribbean, so too will the number of highly vulnerable squatter settlements.

Another constraint to effective planning to reduce the vulnerability of coastal infrastructure and land uses in the Caribbean to climate variability and change, is the prevalence of freehold ownership of land throughout the region. The comprehensiveness of freehold ownership (property rights) makes it difficult to advance a progrmme that restricts land use. This fact will play significantly into any efforts to address impacts of sea level rise. Thus far, charges against non-compliant land-use practices by freehold landowners have been litigated with marginal success.

HUMAN IMPACTS ON ENVIRONMENTAL DEGRADATION

For Caribbean countries to effectively stem the impacts of human activities that exacerbate the effects of sea level rise on coastal ecosystems, they will need to develop a coordinated set of actions linked to projected outcomes. These actions will likely include such measures as targeted regulatory policies; land-use planning; economic and market-based incentives; research, monitoring and mapping; and broadscale efforts for improving public awareness and education about the relationships between human activities and environmental health. While nearly all Caribbean countries have implemented an array of different measures, they are rarely coordinated, or comprehensive or, most problematical, enforced. In other words, they are not part of comprehensive strategic plans. We offer a number of recommendations that Caribbean island nations can evaluate to develop such plans for reversing the impacts of current development practices and protecting coastal resources from future impacts. While these recommendations are generic, they can be adapted by national and local governments.

CLIMATE VARIABILITY AND CHANGE

Studies conducted worldwide have shown that strict adherence to building codes and standards reduce destruction caused by extreme weather events and climate variability. Revised building codes within individual countries could include instructions to :

- Improve construction techniques such as stronger connections (at the ridge board, between the joists and the top plate, between the floor and the foundation, at the foundation footing), long screws/nails, hurricane straps and strong roofing materials;

- Modify engineering designs to include climate change projections, particularly for sea level rise, in addition to historical data typically used;
- Limit the siting of new structures in hazardous areas, restricting siting of any new public buildings in such areas;
- Elevate structures in high hazard areas (*e.g.*, on pilings) through the designation of minimum floor elevations, piling depths and bracing requirements; and
- Add additional specifications to ensure that new buildings are built to better withstand wind and flooding.

It is important to include training progrmmes in code requirements and construction techniques for local builders and carpenters, including the informal housing community. Strengthened building codes are of little use if they are not properly enforced; therefore, any efforts to revise codes or implement new ones should also include training progrmmes for building inspectors.

An important first step towards introducing and enforcing revised building codes in individual Caribbean countries has been the introduction of the Caribbean Uniform Building Code (CUBiC), a regional building standard. In the Eastern Caribbean, a model building code, based on CUBiC, has been developed by the Organization of Eastern Caribbean States to facilitate the introduction of national codes. However, given current climate change projections, the adequacy of CUBiC should be reassessed — efforts should focus on incorporating these regional standards and the specific instructions listed above into each country's regulations and land use plans. Traditionally, in the Caribbean, existing building and planning statutes do not specifically reference building codes; instead, legislation has focused on safeguarding health and property by proper planning and siting.

DEVELOPMENT IN HIGH HAZARD AREAS

In addition to revised building codes for the quality and siting of new structures, regulations should be introduced to plan a strategic retreat of existing development located in low-lying coastal floodplains and high hazard areas along the coast. These regulations should consider:

- Prohibiting the construction of protective structures in sensitive high hazard areas;
- Prohibiting the reconstruction of storm-damaged property in high hazard areas; and
- Conditioning land ownership in high hazard areas to expire when a property owner dies or when sea levels reach a particular point along a map.

Strengthen Regulations to Protect Ecological Buffers

While most Caribbean countries have regulations governing the use of wetlands and mangroves, strengthening these measures to ensure that ecological buffers are protected should be a priority. The filling of wetlands, damming rivers, mining coral and beach sands and cutting of mangroves should be prohibited in order to preserve the natural storm abatement functions of these areas.

LAND PROTECTION TOOLS RECOMMENDATIONS

Develop Comprehensive Land-use Plans

The most immediate need for reducing island vulnerability to sea level rise is development of comprehensive land-use plans that account for high hazard areas and climate risks. Such plans need to be integrated throughout several government agencies and developed through consultative processes to include public participation. Based on hazard maps and projections of climate change, plans could be designed to limit infrastructure and land uses based on vulnerability to both current climatic conditions and projected changes.

Develop and Implement Integrated Coastal Management Plans

While many Caribbean countries have developed or are developing integrated coastal management plans, for example, natural resource management plans that encompass the various sectors impacting the coastal zone, such efforts should be a priority for government agencies charged with land-use planning and natural resource management. Because coastal ecosystems can serve to buffer the impacts of climate variability and change on coastal infrastructure and land uses, national and local level management plans to conserve these ecosystems should be a top priority for adaptation. Ongoing efforts at integrated coastal management should include assessments of the potential impacts of climate change and sea level rise. These plans also provide a framework to balance resource allocation decisions across ministerial lines, because they seek to promote horizontal integration rather than the traditional vertical command and control decision-making process. Similarly, an important aspect of such plans will be the institutionalization of mechanisms for shared management responsibility between government and communities, which will encourage stakeholder participation in decision making processes.

Watershed Management Plans

Many land use decisions in watersheds throughout the Caribbean are not coordinated; as a result, land use practices are often a large source of sedimentation in coastal waters, that limits the capacity of the coastal ecosystems to buffer storm surge and inundation. Land-use planning and natural resource management agencies should be encouraged to work with the various stakeholders in individual watersheds to plan uses that protect riparian forests and agricultural areas and to reduce or control sediment loads to coastal waters. Such plans could include zoning protection for vulnerable watershed areas.

High Hazard Areas along the Coastline

Land-use planning should encourage a strategic retreat from development and infrastructures in low-lying coastal floodplains and high hazard areas; planning and development would be based on retreat plans, where new structures are located on setback lines behind these areas.. Designating a setback line begins with establishing a baseline, which could be drawn along the dune crest.

For armored shorelines, the baseline may be drawn at the theoretical dune crest location — the position where the dune crest was calculated to exist if the shoreline had not been armored. For most erosional beaches with a sand deficit, this theoretical dune crest location may be significantly landward of the seawall or bulkhead. For inlet zones, which are sections of beach in close proximity to tidal inlets, the baseline may be drawn at the most landward shoreline position at any time during the past 30 to 40 years (South Carolina Beachfront Management Act, 1988).

Once the baseline is established a second line of jurisdiction, called the setback line, can be drawn. The setback line is intended to be a projection of where the baseline would be located in 30 to 40 years, landward from the baseline, at a distance equal to the average annual erosion rate multiplied by 30 to 40. For stable or accretion beaches with a zero rate, the setback line should be designed to give property owners reasonable use of their land, while at the same time restricting large commercial structures from the beach.

Integrate Regional Disaster Mitigation Strategies

Historically development of physical and disaster mitigation plans has been done separately under different ministries with little or no integration. This also has been true in the United States: while the coastal management agency is commonly located in either natural resource management departments or the land-use planning agencies such as the state planning office, the state emergency management agency holds the lead for disaster response. Consequently, coordination and integration of these functions has developed slowly. However, such coordination of functions has facilitated improvements in planning and development by taking into consideration the need for improved building codes to reduce the loss of life and property; discouragement of development in high hazard areas, incorporation of a retreat policy and promotion of the use of soft structures to control erosion and flooding.

Encourage Retreat from High Hazard Areas

Caribbean countries attempting to maintain ecological buffers against extreme weather events or to encourage retreat approaches from high hazard areas may wish to explore one or more of the following land protection tools. The most effective approach involves using a combination of tools based on the unique environmental and development conditions on each island. All of these tools should fall within the framework established under a comprehensive land use plan. Land protection tools can then complement zoning and building codes to carryout the goals of the comprehensive plan.

Donated Conservation Easements. Legal agreements undertaken between a landowner and a land trust or government agency that allows landowners to permanently limit or prohibit development on their property. Conservation easements become part of the property title and hence the easement transfers with any change in property ownership.

Purchased Development Rights (PDRs). Voluntary legal agreements that allow owners to sell the rights to develop their properties to government agencies, or non-profit organizations. As is the case with donated conservation easements, a conservation easement is then placed on the property and recorded as part of the property title. Typically PDRs are used to limit the future use of the property to agricultural, forestry or open space. For example, a government agency might identify several parcels of undeveloped coastal lands in vulnerable areas that serve as buffers between settlements and floodwaters. The agency might then want to approach property owners to purchase the development rights and then limit the future use to open space.

Transferred Development Rights (TDRs). Established by local ordinances that create "sending areas" (preservation areas) and "receiving areas" (areas where growth and development are encouraged). Property owners in the sending area receive development right credits, which they can sell in exchange for not developing their property. Real estate developers, speculators, or government agencies may purchase the development right credits and use them to increase existing or planned development in receiving areas.

Land Acquisition. Allows property owners to sell or donate their property to a public agency or land conservation organization,giving the public agency full control over future use to include development or preservation. Government could acquire high hazard areas and convert the lands into natural preserves as ecological buffers. There is a growing trend towards such acquisition in many countries, as it greatly reduces the vulnerability of coastal areas to extreme weather events and the damages associated with such events. In the U.S., for example, some areas have held public referendums to establish a small "bed tax" to be paid by every tourist staying overnight in hotels, condos, etc. Proceeds from this tax are used to create a land acquisition fund to acquire vulnerable hazard areas and restore their natural buffering capacity. This tool should be used to acquire, restore and preserve wetland and mangrove systems, as well as dune systems that can act as ecological buffers against storm surge and inundation.

Retreat and Accommodation

For coastal urban areas with high population densities, retreat from, or accommodation of, sea level rise may not be an option. In such cases, hard structures such as sea walls may need to be designed to allow for future modification to ameliorate sea level rise. Integral to the design of any hard structure, a sand management study should be initiated to assess the impact on beach sand dynamics. It should be noted that hard structures are heavily capital intensive, which may deter their widespread use. Thus, consideration of hard structures should be limited only to those areas where capital development has occurred and the value of existing infrastructure far outweighs the cost of the hard structure approach.

ECONOMIC AND MARKET-BASED INCENTIVES RECOMMENDATIONS

There is a direct linkage between the declining health of coastal ecosystems and the growth of the tourism industry in the Caribbean. The demands on natural resources to service the expanding tourism industry is having detrimental effects on coastal ecosystems, rendering these systems more vulnerable to climate change and reducing their capacity to buffer coastal infrastructure from extreme weather events and floods. To combat these trends, market-based incentives are already being used by Caribbean nations to promote a more sustainable tourism industry (Jamaica Sustainable Development Networking Programme). These incentives are wide-ranging and include:

- Duty and tariff relief and/or tax deductions for environmental resource protection and water-conservation equipment;
- Water and sewage charges for both industrial and domestic users;
- Private sector initiatives such as the Blue Flag Campaign or Green Globe 21 could be used to encourage eco-tourism and environmental resource preservation. The Blue Flag is an eco-label awarded by a non-profit organization to beaches and marinas in countries around the world, based on environmental management criteria. Similarly, Green Globe 21 is a collective body of roughly 1000 travel and tourism companies in over 100 countries, who have committed to specific environmental management standards; and
- Green award progrmmes for hotels that display the best conservation practices or education progrmme.

Link Property Insurance with Construction Quality

The relatively poor quality of many built structures in the coastal zone have increased the vulnerability of these areas to climate variability and sea level rise. Perhaps no market-based tool can be as effective in improving the quality of such structures as insurance incentives. These incentives would include lower deductibles and/or premiums for those who invest in adaptation measures, such as homeowners and businesses who alter or upgrade properties to better withstand hurricanes-force winds.

Reduced insurance premiums for policyholders to make their homes and businesses more hurricane-resistant could be based on available information, such as "Making Your Home Hurricane Resistant" and the "Professional Guide to Performance-Based Design Upgrade for Hurricane Resistant Construction," produced by an engineering firm in Barbados.

One key to such incentives is that the insurer is able to ensure the quality of the retrofit work that is done. However, such retrofit work and adaptation measures should benefit the insurance company. After a sufficient number of property owners at risk in the company's portfolio undertake adaptation measures, the company can expect its aggregate catastrophe Probable Maximum Loss (PML) to be lower. The primary insurer should be able to negotiate lower rates from the reinsurer for this effort, which combined with lower incurred claims on the retained risk should more than offset the reduced premium income.

Promote Development in High Hazard Areas

In many cases, national governments reallocate economic development loans in order to reconstruct and repair structures damaged from extreme weather events.

This practice can in fact be considered a government subsidy, which on a long-term basis encourages development and land uses in vulnerable areas, as owners know that government will help repair any damages caused by a hurricane or a similar event. Over the long-term,this practice disrupts national economic development progrmmes, and reinforces a state of underdevelopment in many vulnerable island and low-lying Caribbean states. Governments are forced to borrow more funds from external donors for reconstruction and repairs following an extreme weather event, diverting funds that might have been used on national education progrmmes, for example. This is not to suggest that some level of disaster relief is not necessary after an extreme weather event, but continued reconstruction of buildings and structures that are located in highly vulnerable areas can be considered a subsidy for land uses susceptible to climate variability and change.

Establish a Revolving Loan Fund for Home Improvement

Because many of the most vulnerable settlements in coastal zones are low-income, informal housing, countries could encourage the establishment of revolving loan funds to provide families with small loans for home improvement and hurricane-resistant retrofitting.

Improve Public Awareness

Much lip service is often given to the importance of public understanding and education in public policy issues. But public awareness and knowledge is critical in addressing the impacts of climate change and variability and changing attitudes. Governments must embark on a sustained education campaign that is part of a broadscale communications plan designed to demonstrate that key public constituencies understand the impacts of climate variability and change on vulnerable areas, as well as potential adaptation measures they can take. A knowledgeable consumer, for example, should be reluctant to purchase property in high hazard areas. Dissemination of targeted information also helps to ensure that various market sectors take into consideration potential impacts as well as new market growth opportunities. Central to any education progrmme must include the following: promoting awareness of climate change impacts; communicating information about vulnerable and hazardous areas; educating key stakeholders on tools and practices that can be applied to reduce impacts and encouraging participation at all levels in decision-making.

Continue to Build and Expand Long-term Beach Monitoring Progrmmes

Through the CPACC project, several countries throughout the region have begun to implement beach-monitoring progrmmes at selected beaches, thereby establishing a series of benchmarks. In existing and future monitoring sites each benchmark, a metal disk set in concrete, has a known vertical elevation that is used as the starting point for the beach survey. In addition, the horizontal coordinates of each benchmark are established, so the benchmark can be replaced in the same location if it is ever destroyed. The information from beach-monitoring progrmmes should be increasingly used in coastal land-use plans and retreat progrmmes, in order to delineate the position of a baseline, set at either the actual dune crest for natural beaches or the theoretical dune crest for armored beaches.

Promote Increased use of GIS and Remote-sensing

Geographic information systems and databases should be promoted to provide resource managers with information on the state of resources and to assess their vulnerability to climate change.

Inventories that quantify and qualify current uses of land, distribution and species diversity provide the baseline information managers need to prepare development scenarios and strategies to assess vulnerabilities to climate change, projected impacts on ecosystems and land-use patterns, and potential land use conflicts. Key to producing and maintaining this information is the development of a local capacity and skilled workforce to design and implement these databases.

Climate Change Monitoring Institutions

A key deficiency in current climate monitoring is the lack of a capability to provide timely observations to both regional and national forecasters, observers, resource managers and governmental sectors. Coupled with the development of an information-sharing network is the development of databases generated from national and regional projects, observations and other complementary information such as growth and development trends. It should also be noted that paramount to any of the aforementioned actions is the development of national and regional capabilities for display, analysis and reporting of data and information related to climate variability and change. This "local capacity" varies widely from island to island as do hardware and software capabilities. Development of regional or national training centers would be a first step toward promoting growth and development of data and information capabilities among small island nations.

Hazard Mapping of Coastal Zones

A significant need is the identification of areas vulnerable to sea level rise. Government agencies and research institutions should prioritize capacity building for vulnerability assessments to delineate critical hazard areas, particularly as a result of current and projected storm surge and wave action. Whether governments or research institutions perform the mapping, the results can then be integrated into the land-use planning and development process to allow agencies to identify areas for retreat strategies and many of the land protection tools described previously. This process should be coupled with development of data and information clearinghouses to facilitate the exchange of information at local, national and regional levels. In particular, hazard maps based on current and future vulnerability should be a major feature of any education and information progrmme to coastal residents and stakeholders.

4

Climate Variability and Projections

RELEVANT CLIMATE PARAMETERS

Climate in a narrow sense is usually defined as the average weather, or more rigorously, as the statistical description in terms of the mean and variability of relevant quantities over a period of time ranging from months to thousands or millions of years. The classical period for averaging these variables is 30 years, as defined by the World Meteorological Organization. The relevant climate parameters include temperature, precipitation and wind. Climate in a wider sense is the state, including a statistical description, of the climate system. The annual rainfall in the State varies roughly from 50 to 350 cm. In the districts of Bijapur, Raichur, Bellary and southern half of Gulbarga, the rainfall is lowest varying from 50 to 60 cm. The rainfall increases significantly in the western part of the State and reaches its maximum over the coastal belt. The south-west monsoon is the principal rainy season during which the State receives 80% of its rainfall. Rainfall in the winter season (January to February) is less than one per cent of the annual total, in the hot pre-monsoon season (March to May) about 7% and in the post-monsoon season about 12%.

Tropical monsoon climate covers the entire coastal belt and adjoining areas. The climate in this region is hot with excessive rainfall during the monsoon season *i.e.*, June to September. The southern half of the State experiences hot and seasonally dry tropical savana climate while most of the northern half experiences hot, semi-arid and tropical steppe type of climate. The climate of the State varies with the seasons. The mild winter season from January to February is followed by relative warmer summer season from March to May. The period from October to December forms the post-monsoon or northeast monsoon season. The period from October to March, covering the post-monsoon and winter seasons, is generally pleasant over the entire State except during a few spells of rain associated with north-east monsoon which affects the south-eastern parts of the State during October to December.

The months April and May are hot, very dry and generally uncomfortable. Weather tends to be oppressive during June due to high humidity and temperature. The next three months (July, August and September) are comfortable due to reduced day temperature although the humidity continues to be very high.

CURRENT CLIMATE VARIABILITY

The starting point for an investigation of adaptation to future climate change is to develop an understanding on the adaptation to current climate variability. This means that the focus for the investigation of adaptation to climate change should not be based only on scenarios of the future but also on the analysis of present vulnerability in the face of current climate variability. Rainfall and Temperature are subjected to variability on all time scales: intra-seasonal, inter-annual, decadal, centennial, etc.

Food production system, water availability, water resources etc. are sensitive to intra-seasonal, inter-annual, decadal variability in these two important climate variables. Global warming and Climate Change is projected to increase the number of extreme temperature and rainfall events, and hence climate variability is expected to show an upward trend. It is very important to understand the past trends and variability in rainfall, minimum and maximum temperature in Karnataka since the knowledge on the past could provide guidance for the future.

Climate variability refers to variations in the mean state (of temperature, monthly rainfall, etc.) and other statistics (such as standard deviations, statistics of extremes, etc.) of the climate on all temporal and spatial scales beyond that of individual weather events. Variability may be due to natural internal processes within the climate system (internal variability), or to variations in natural (*e.g.* solar and volcanic) external forcing (external variability).

We focus on the current mean climate and climate variability in Karnataka at district level and investigate how changes in them will alter Karnataka's vulnerability to climate change. Precipitation and temperature are used as the key climate variables in this analysis.

DATA AND METHODOLOGY

In order to study the climate variability, 3 meteorological sub-divisions in Karnataka were selected.

- Coastal Karnataka (Dakshina Kannada and Uttara Kannada districts),
- North Interior Karnataka (Belgaum, Bidar, Bijapur, Dharwad, Gulbarga and Raichur districts) and
- South Interior Karnataka (the remaining districts of Bangalore Rural, Bangalore, Bellary, Chikmagalur, Chitradurga, Kodagu, Hassan, Kolar, Mysore, Mandya, Shimoga and Tumkur districts)

The study considers the climate variability only for southwest monsoon (rainy) season. The high resolution (0.5° x 0.5° lat. and long.) daily gridded rainfall dataset for a period of 35 years (1971–2005) provided by Indian Meteorological Department (IMD) for precipitation and the Climatic Research Unit Time Series (CRU TS) version 2.10 on a 0.5° x 0.5° lat. and long. resolution monthly dataset spanning 102 years (1901-2002) for temperature are used. District-wise data is obtained by re-gridding the dataset to 0.1° lat. x 0.1° long, and re-aggregating by districts to study the climate variability at district level.

Rainfall Variability

Information on spatial and temporal variations of rainfall is important in understanding the hydrological balance on a global/regional scale. The distribution of precipitation is also important for water management in agriculture, power generation and drought-monitoring. In India, rainfall received during the southwest monsoon season (June–September) is crucial for its economy. Real-time monitoring of rainfall distribution on a daily basis is required to evaluate the progress and status of monsoon and to initiate necessary action to control drought/flood situations.

The spatial distribution of the mean monsoon (June–September) rainfall averaged for the period 1971–2005 derived from the IMD gridded rainfall dataset.

- The highest climatological rainfall is seen over Costal Karnataka districts (Dakshina Kannada, Udupi and Uttara Kannada) (>25 mm/day) and parts of South Interior Karnataka districts (Kodagu, Hassan and Shimoga) (>12 mm/day).
- Most of the South Interior Karnataka districts and North Interior Karnataka districts of this region experience lowest (<5mm/day) seasonal mean rainfall in a climatological sense.

A normal monsoon with an evenly distributed rainfall throughout the region is desirable, while an extreme event of flood or drought over the region constitutes a natural hazard. Variation in seasonal monsoon rainfall may be considered as a metric to examine climate variability/change over the Karnataka monsoon domain. The year-to-year variability in monsoon rainfall shows extreme hydrological events (large-scale droughts and floods) that could be resulting in reduction in agricultural output and affecting the vast population and economy.

The dotted lines indicate the value of the percent of rainfall corresponding to one standard deviation, which is 14.1% of the mean. Drought (Flood) years can be defined as the years when the precipitation is lower (higher) than 14.1% of the mean.

Coefficient of Variation is defined as the inter-annual variability (estimated as the standard deviation) of rainfall over the region as a fraction of mean. Higher values of C.V. indicate larger inter-annual variability and vice versa.

- The coefficient of variation of rainfall is higher over South Interior districts (>55%) of Karnataka.
- The variability is very high in a couple of districts (Chikballapur, Chitradurga, Gadag, Kolar, Mandya and Tumkur) of Karnataka, where the highest C.V. (>85%) is observed.

Standard deviations of rainfall for the districts. We have also prepared a time series of precipitation rate averaged over the area 11.5–18.5° N, 74–78.5° E (Karnataka Region) for a period of 54 years (1951-2004) using the IMD gridded data to examine any significant trends.

The linear trend in the dataset, calculated using a least mean square method is shown by the straight line. The slope of the straight line is -1.0 mm/ day per 100 yr (about 6% decline in 50 years) which clearly indicates a decrease in the precipitation rate over Karnataka. Figure clearly reflects a decrease in precipitation trend in Coastal Karnataka and North Interior Karnataka meteorological subdivisions.

TEMPERATURE VARIABILITY

The meteorological measurements of temperature for Karnataka are analyzed.

The district-wise variation of the annual mean minimum and maximum temperature averaged for the period 1901–2002 derived from CRU-TS dataset.

- The highest annual mean minimum temperature (>=21 °C) is observed over Raichur, Udupi, Gulbarga and Yadgir district.
- Maximum temperature (>33 °C) is noticed in most of the North Interior Karnataka districts - Raichur, Bijapur, Gulbarga and Yadgir.

District-wise Coefficient of Variation (C.V) for annual mean minimum and maximum temperature. Higher values of C.V. indicate larger inter-annual variability and vice versa. C.V of annual mean minimum and maximum temperature for districts is listed in Tables respectively.

- The coefficient of variation of annual mean minimum temperature is very low over Coastal Karnataka districts (< 8.5%).
- The variability is very high in a couple of districts (Bidar and Gulbarga) of North Interior Karnataka, where the highest C.V. (>14.5%) is observed.
- Majority of the districts in South-east Karnataka as well as that of North Interior Karnataka show a high variability (>9.0%) for the annual mean maximum temperature.

DISTRICT-WISE TREND ANALYSIS OF MINIMUM AND MAXIMUM TEMPERATURE

The analysis of the meteorological measurements of temperature for Karnataka shows a steady warming trend in both the minimum and maximum temperatures. Both day and night temperatures are more or less uniform over the State, except at the coastal region and high elevated plateau.

- The spatial pattern of minimum temperature trend, an increase all over the region. Districts of North Interior Karnataka (Bidar, Bijapur, Gulbarga, Yadgir and Raichur) experience an increase in minimum temperature of e" 0.6°C in the last 100 years.
- The maximum temperature trend Figure indicates an increase of e" 0.6°C in the last 100 years over Gulbarga, Bijapur, Gulbarga, Bidar, Raichur and Yadgir.

FUTURE CLIMATE PROJECTIONS

Climate change projections for any geographic region are obtained using GCMs (global circulation models) and RCMs (regional climate models). Normally, climate change projections are made for variables such as:

- Surface air temperature (height of 1.5m) – Average, maximum and minimum
- Precipitation (both snow and rainfall)
- Relative humidity
- Radiation (both shortwave and longwave)
- Wind speed (height of 1.5m)

Model and Methods

For climate change projections, we have used simulation data from the global climate model, HadCM3 from the Hadley Centre, UK. HadCM3 has been used recently for generating climate change projections for various parts of the Indian subcontinent. According to that study, the Indian subcontinent could be subjected to an average of over 4° C increase in temperature by 2085 for SRES A2 scenario.

GCM and SRES scenario used: In this report, data from the HadCM3 global climate model downscaled by PRECIS model, a regional climate model for downscaling climate projections, is used. The combination of HadCM3 and PRECIS models is known as the HadRM3 model. The pathways for atmospheric greenhouse gases (*e.g.* CO_2, CH_4, N_2O, CFCs) are prescribed from the SRES A1B mid-term (2021-2050) projections. Climate change projections were made:

- For daily values of temperature (average, maximum, minimum)
- For daily values of precipitation
- At grid-spacing of 0.44250 latitude by 0.44250 longitude
- For periods of 2021-2050 relative to the baseline period 1961-1990 (also referred to as either 1975 or 1970s)

Derivation of district-wise data: Data derived from the PRECIS model outputs (which had a grid spacing of 0.4425° latitude by 0.4425° longitude) was regridded to 0.2° in latitude and 0.2° in longitude. This ensures that enough grids fall inside each district. Then, the data was re-aggregated (as averages) at the district-level.

PROJECTIONS FOR THE FREQUENCY OF DROUGHTS

We define a period of absence of rainfall (daily rainfall < 2.5 mm) for 40 or more contiguous days as "an incidence of severe drought". This agrees with the FAO definition of droughts. Further, we identify two major growing seasons in Karnataka: *Kharif* (July 2nd week to October 2nd week) and *Rabi* (September 1st week to February 4th week). The number of incidences of severe droughts are estimated for 2021-2050 for each grid point and compared to the baseline. The percentage increase is calculated; this increase gives an indication of possible increases in drought instances for that grid-point. Grids that are not drought-prone (number of severe drought instances less than 20 in the 30-year baseline) are excluded.

We find that:

- In the Kharif season, most northern districts are projected to have an increase in drought incidences by10-80%
- Districts of Koppal and Yadgir are projected to have almost a doubling of drought frequency in the Kharif season
- In the Rabi season, drought frequency is projected to increase in most of the eastern districts of the state.
- The western parts of the state are projected to have more rainfall and hence less number of droughts in the Rabi season.

Multi-GCM Analysis of Temperature and Rainfall Projections

Method: To ensure reliability of projections, analyses of multiple models are often used to build confidence. We also use a similar multi-model methodology; that is, we analyze how much the projection from other GCMs agree with that of HadRM3 (from which all our district-wise projections have been derived) for A1B scenario. All GCM grids falling in the general rectangular region containing Karnataka (latitude between 11.0 ° N and 18.0 ° N and longitude between 74.0 ° E to 78.0 ° E) were analyzed.

Temperature: Figures show the increase in temperature projected by different GCMs. Mean of all GCMs is indicated by the dotted lines. One can observe that the increases projected by HadCM3 are quite near the mean projections.

Rainfall: A similar comparison, but for annual increase in rainfall predicted by various models. HadCM3 projections are quite near the average projections except for the period of 2065-2080.

Hence, we find that HadRM3 temperature and rainfall projections (which are based on HadCM3) are nearly representative of the mean of IPCC models, especially in the near future (2021-2050). Also, the high degree of agreement among GCMs with respect to temperature projections and less agreement with respect to rainfall projections can be observed. However, we caution that the spatial pattern of projection from other GCMs could be different. The signal to noise ratio is also usually higher at local scales. To address the uncertainty at local district levels, downscaled data from multiple GCMs will have to be analyzed in the future.

COMPARISON OF PROJECTED CHANGES IN RAINFALL WITH OBSERVED CHANGES

We compare the changes in rainfall (quantum of increase/decrease) that is projected by the HadCM3 model (over the period of 1975 to 2035) against trends that are already observed over the period of 1971-2005. For example, HadCM3 could be projecting that the annual rainfall would increase at a rate of 3% per decade (from 1975 to 2035) for a particular district. The question is: what is the actual observed increase seen for that district? To address this question, we compare the projected trend and the observed trend for the districts. The method used for these is briefly outlined below.

Method

Projected trend: (a) The average rainfall for the baseline period of 1961-1990 (A1B) is obtained from the HadCM3 model; (b) The average rainfall for the period of 2021-2050 (A1B) was obtained from the HadCM3 model; c) The percentage increase between (a) and (b) are computed. The data was then scaled so that the percent increase is over a 10-year period.

Observed trend: a) The climatological mean rainfall for the period of 1971-1991 was obtained from 0.5 degree x 0.5 degree IMD data; b) The climatological mean rainfall for the period of 1985-2005 was obtained from 0.5 degree x 0.5 degree IMD data; c) The percentage increase between (a) and (b) were computed. The data was then scaled so that the percent increase is over a 10-year period.

Correlation between Projected and Observed Rainfall

Correlation is high between model projections and actual observed rainfall trends for the annual-mean and the post monsoon (OND) period. Therefore, we may conclude that projections by the model for annual and OND rainfall are quite robust. Hence, it can be stated with some confidence that: 1) Annual rainfall would generally decrease in the North-east and South-west parts of the state; 2) The OND rainfall would decrease in the South-west part of the state.

Correlations are weak for the summer monsoon (JJAS) period. Observed trends mostly have a much higher magnitude (of the order of 10-20%) than projections (of the order of 0-5%). Therefore, our projections may be underestimating the trend.

5

Vegetation of the Forest Zones

FOREST ZONES

Although Ceylon lies entirely south of the Tropic of Cancer, no place in the island is more than seventy miles from the ocean and the temperatures are correspondingly moderate. The monsoons (northeast from November to February and southwest from May to October) and a hilly relief combine to create diverse climatic conditions that give rise to an exceptionally rich variety of plant communities. Between Colombo on the west coast and Batticaloa, 140 miles away on the east coast, there are no fewer than six main climatic and corresponding forest zones.

THE TROPICAL EVERGREEN ZONE

The southwestern part of the island is covered by tropical evergreen rain forest similar to that usually found in warm tropical regions of heavy rainfall and high humidity. This region benefits from both monsoons and receives a well-distributed rainfall of over 100 inches per annum; the mean annual temperature is approximately 80° F., with small diurnal and monthly ranges.

The vegetation is luxuriant and trees usually attain a height of more than 100 feet; climbers abound. The forest growth consists of a large variety of species producing timber mainly of nondurable quality. Competition between species is so great that it is not easy to select any one as being typical of an area. In certain localities Dipterocarpus zeylanicus produces almost pure stands. Other strong competitors are Doona congestiflora (thiniya), Cyathocalyx zeylanicus (kekala), Calophyllum bracteatum (kina), Artocarpus nobilis (del), Cinnamomum multiflorum (wal-kurundu), Ficus spp., Melia composita (lunumidella), Palaquium grande (mihiriya) and Pericopsis mooniana (nedun).

THE WET MONTANE. ZONE

Northeast of the tropical evergreen forest zone lies the wet montane zone.

A steady decrease in the luxuriance of the tree growth with altitude is observed. The species found at lower elevations begin to disappear and others take their place. The more prominent among the latter are Michelia champaca (sapu), Syzigium umbrosum (damba), Acronychia pedunculata (ankenda) and Kurrimia zeylanica (palen). Above about 4,500 feet, the forest becomes a true montane evergreen forest with species such as Syzigium umbrosum, Syzigium rotundifolium,Gordonia zeylanica, Meliosma simplicifolia (elbedda) and Calophyllum walkeri (kina). Higher still the trees are even more stunted and branched. Dead and dying trees are seen practically everywhere.

Although the annual rainfall is over 100 inches and is well distributed throughout the year, strong winds and very low temperatures from December to February (26.1°F. recorded in February, 1914) prevent good forest development. This type of forest alternates with undulating grasslands locally called patanas, on which the only tree species is the sparsely distributed Rhododendron arboreum (ma-ratmal). The shifting cultivators who originally cleared the land of its forest cover, together with the annual grass fires that followed, may be responsible for the almost total absence of forest in these areas.

THE INTERMEDIATE MONTANE ZONE

East of the wet montane zone is an area of lower rainfall (about 75 inches yearly), shallow soil of low storage capacity and droughts of four months or longer in duration-conditions not conducive to the development of trees. The montane evergreen forest of this intermediate zone is confined to watercourses, folds in the hills and other places with a moist soil and consists of a mixture of species found at both higher and lower altitudes. Below 3,000 feet occur species characteristic of the adjacent savanna forest of the lower elevations, such as Terminalia chebula (aralu), Terminalia belerica (bulu), Emblica officinalis (nelli), Anogeissus latifolia (dawu), Butea monosperma (gaskela) and Pterocarpus marsupium (gammalu).

Lantana aculeata, originally introduced into Ceylon as a flower plant, has, in many places, become a persistent weed. The large extent of grassland (dry patanas) that alternates with these patches of forest is due to the enormous runoff over steep slopes, to long periods of drought and to the shallowness of the soil above the gneiss rock, although in some localities shifting cultivation and subsequent grass fires have been responsible. The only tree species which survives in these dry patanas is Careya coccinea.

THE SAVANNA FOREST ZONE

At approximately 1,500 feet one enters the parklike savanna forest, locally called damana or, in the wet zone, talawa, an apparently unending carpet of grass with a few scattered trees. The annual rainfall here is more than 75 inches, but the period of drought in some years lasts as long as five months.

This country of indescribable beauty is still the haunt of the elephant, buffalo and leopard. A paradise for the hunter, landscape painter and poet, it is of no economic value as a forest. There is a total absence of undergrowth and the herb layer consists of grasses, of which Imperata cylindrica (illuk) is the most prominent. Shallow, poor soil, long periods of drought and the destructive activity of man have contributed to the making of this plant formation.

THE SEMI-EVERGREEN MONSOON DRY FOREST ZONE

This lies east of the savanna forest zone and has a similar climate. Most of the trees are deciduous and their average height is about 60 feet. These forests produce some of the best timbers in the world -- Chloroxylon swietenia (satinwood), Berrya cordifolia (Trincomalee wood, halmilla), Vitex pinnata (milla), Diospyros ebenum (ebony), Alseodaphne semecarpifolia (ranai) and Manilkara hexandra (palu). These valuable species grow in mixed stands with others less valuable, such as Hemicyclia sepiaria (wira), Ficus spp., Walsura piscidia (kiri-kon), Pleurostylia opposita (panaka) and Aglaia roxburghiana (kannakompu).

THE COASTAL JUNGLE ZONE

The arid coastal belt in the east, northwest and southeast of the island is covered by a thorny scrub jungle, which does not exceed 25 feet in height. This zone receives less than 50 inches of rain concentrated in about four months of the year. The air humidity ranges from 60 to 75 per cent. The plant community consists of shrubs such as Gmelina arborca (et-demata), Randia dumetorum (kukuruman) and Memecylon rostratum (kuratiya). The scattered tree species found here are Cassia fistula (ehela), Calophyllum calaba (gurukina), Manilkara hexandra (palu), Hemicyclia sepiaria (wira), Euphorbia antiquorum (daluk) and Chloroxylon swietenia (satinwood). Commercially these forests are of no value, but merely provide cover for the poor sandy soil.

Since climatic changes are not abrupt there are few sharp divisions in the forest vegetation: the zones merge into one another, forming intermediate communities. The area between the tropical evergreen and the semi-evergreen monsoon dry forest is covered by a forest that has some of the characteristics of both, but most of this plant formation still awaits proper scientific description.

MANGROVES VEGETATION

Mangrove vegetation is the constant associate of the shallow ocean shores, mudflats and river embayments of almost all of the high islands throughout Oceania. Only in Hawaii -- most distant from sources of dispersal -- were mangroves absent until relatively recent times. Yet they have been introduced here and are now developing semi-natural stands. Mangrove sites are rare on low islands. As one progresses from open sea water landward to upland and also to brackish water, gradual changes occur in the vegetation, which is generally segregated in definite belts. The forest fronting the ocean -- usually much the largest in acreage -- is dominated by Rhizophora spp. Frequently just back of the front Bruguiera spp. are common. Farther landward are large numbers of other trees: Xylocarpus (carapa) spp., Sonneratia caseolaris, Heritiera littoralis, Avicennia marina, Lumnitzera littorea, Excoccaria agallocha and the colonial 2-meter-tall fern, Acrostichum aureum. The important thatch palm "nipa" (Nipa fruticans) is predominant in brackish waters.

Economic utilization of this vegetation is most highly developed in Fiji, where 52 square miles (33,000 acres), owned by the Crown, are commercially worked. On the basis of a mean annual increment of 130 to 150 cubic feet per acre, the lands are handled for fuel wood on a 40-year rotation, with Bruguiera and Rhizophora supplying 95 per cent of the take.

THE XEROTROPICAL ZONE

Dry Deserts. No survey has been made of the islands to locate those areas within the xerotropical zone where the precipitation is so light (certainly no More than 15 inches) that they have either a distinctive desert flora, or a permanently nonforest vegetation due to other causes than human interference, or both. A few small areas in New Guinea are reportedly in this category and also a few lowlands in the Hawaiian islands, as well as some of the highest summits above the cloud zone but below the frost line. At these high elevations in Hawaii, however, where there are scattered dwarf outliers of the Sophora chrysophylla forest, the vegetation of pre-goat days has not been satisfactorily interpreted.

Some low coral islands, such as Jarvis, may likewise belong in this class. In these cases, however, the few plants present are those of the cosmopolitan strand flora and no distinctive desert vegetation is apparent.

It is important to distinguish between the dry deserts, which are correlated with the "natural" or nonhuman environment and certain large areas that now resemble such deserts because of the direct and indirect effects of man. The former are relatively unalterable, but the latter, at least in some instances, could be converted into forests and even into productive watersheds under long-range management.

Dry Forest, Scrub and Savanna. A zone of dry forest, scrub and savanna is generally found where there is less than 50 to 60 inches of rain per year. On typical oceanic volcanic islands it occurs at low elevations, especially on the lee sides. On very high volcanic islands it may again appear above the mossy forest, although such occurrences have not been adequately described. On continental islands it may occur at any elevation where the precipitation is low. Very few detailed accounts of this vegetation exist, for understandable reasons. In its original form it has been totally destroyed by the natives and hordes of recent introductions have filled the vacuum. Since such areas seemingly bear only worthless herbs and shrubs, grazing rather than forestry has been their hoped-for use. Nevertheless, in all foot slopes in this zone in the Hawaiian islands the small tree Leucaena glauca can invade and replace various shrubs such as the ubiquitous Acacia farnesiana and, in turn, is locally invaded by still other tree species. Coastal plains have, within the past half century, developed semi-natural forests of Prosopis chilensis, 40 feet high, that are of economic value not only for firewood but for cattle shade, cattle forage (from fruits) and apiculture. It may well be that other valuable trees could be developed in this zone.

In New Guinea areas apparently comparable bear "savanna forest" The grasses are typically Imperata cylindrica (arundinacea) and Saccharum spontaneum. The scattered trees, growing to a maximum height of 50 feet, consist, at least in Papua, of several species of eucalyptus, including E. alba, E. papuana, E. clavigera and E. tereticornis, while elsewhere Melaleuca spp. and Casuarina spp. are common. As a source of fuel wood and post wood, these lands have value for both aborigines and European settlers. What other trees they might grow under fire management is totally unknown, for no relict unburned area has been described.

THE CLOUD ZONE

The vegetation of this zone is sometimes, but not always, separated into two relatively pronounced forest belts, especially in the continental islands.

New Guinea has a cloud zone with a markedly distinct flora of highaltitude continental origins. The lower belt ("mid-mountain forests") begins suddenly at the lower edge of the clouds, which may be as low as 2,000 feet, or as high as 6,000. It is the region with timber of highest commercial value, though fires set by the aborigines are destroying much of it, converting large tracts into savanna. Araucaria cunninghamii is one of the most striking, abundant and valuable of the trees. In addition, there are many oaks with Nothofagus just above. Eugenia, Cryptocarya, Sideroxylon and Podocarpus are among other trees of this belt.

The mossy forest, which in New Guinea begins between 5,500 and 8,000 feet, has one of the most distinctive appearances of all the world's vegetation. Here the trees seldom exceed 25 feet in height. Mosses and lichens cover and hang from everything, bloating the smallest twigs to ten or twenty times their diameter. The air is cloud-saturated, the moss water-saturated.

In New Guinea, Lane-Poole noted Eugenia and Podocarpus as most abundant. At one time, he observed a morning drought, when all the moss was brittle and exploded to dust at the touch, a condition that may help to explain the large areas of burnt mossy forest he had seen. Thus, even here native fires have their influence. On Fiji the mossy forest begins at about 3,000 feet. In Micronesia it is found only on Ponape and Kusaie, beginning at about

1,350 and 1,450 feet respectively; here the most abundant woody plants are Alsophila ponapeana, Gynotroches axillaris, Syzygium carolinense and Astronidium ponapense. Samoa has a comparable type that begins at about 1,500 feet. In the Hawaiian Islands this belt, sometimes suffering from pig destruction, may start as low as 1,800 feet. The predominant woody plants, rarely different from those of the forest below, include Metrosideros, Bobea, Straussia, Fagara and Labordia.

On Maui and Hawaii islands, both the cloud zone and the forest just below are squeezed out of existence on the dry, leeward side and the dry forest that would otherwise lie below them ascends to higher elevations.

From the foregoing it can be seen that the mossy forest is highly variable in both altitude and floristic composition. Some of the variation in floristic composition is but a part of the floristic individuality of each island. The evidence is not yet conclusive as to whether the cloud zone tends to possess its own distinctive flora, or is a reduction in tree growth and a lush growth of mosses that can be superposed on any tropical flora under conditions of excessive cloudiness.

THE PLUVIOTROPICAL ZONE

This zone of pluviotropical vegetation, areally predominant on the rainy Solomons and New Hebrides and occurring on the intermediate slopes of almost all the high islands, continental and oceanic, includes what is popularly known as "jungle" (although that Hindu term in its original sense referred to secondary thickets on uncultivated land, perhaps even in arid zones). It embraces not only the rain forests, that profuse expression of tropical luxuriance made famous by Tarzan, but also much land now covered with savanna or with thickets of worthless shrubs.

In New Guinea burned savannas mixed with secondary forest extend completely through this zone to the lower limit of the cloud zone at 5,500 feet, with Casuarina nodiflora common at the higher elevations. Such a mosaic of savanna and forest has much in common with the Parang vegetation of the Philippines, where the same species of Imperata and Saccharurn are also dominant in the grassland phase. In New Guinea Octomeles sumatrana, an aggressive, valuable, 150-foot-tall floodplain tree, forms one of the few pure forest types in this part of the tropics. Nevertheless, it is all but impossible to pick any typical genera of the mixed forest. Castanopsis junghuhnii has attracted attention because of its frequency and its habit of occurring in pure patches of an acre or more. A few dipterocarps are present, but they never predominate as they do in the Philippines and Southeast Asia.

In New Caledonia savannas similar to those of New Guinea are widespread, but complex rain forests predominate throughout the island. Melaleuca leucadendron is a typical tree. Related species of Araucaria and Agathis occur at higher elevations.

Fiji has its talasinga (sun-burnt lands), rolling plains extending up into the mountains to altitudes of 1,500 feet. The scattered woody plants include Acacia richii, Casuarina equisetifolia, Dodonaea viscosa and Psidium guayava. The rain forests of these islands, from sea level to 3,000 feet, are complex floristically, but much less so than those of New Guinea. In Tahiti most of the vegetation is of the pluviotropical type and is closely related to that of other Polynesian high islands. Stenolobium stans, of Atlantic tropical America, is said to be on the march here, with armies of young trees ascending these slopes"

In Hawaii the forests of the lowest portions of this zone were destroyed either before white men came or within their memory. Vast acreages are now covered with lantana, guava and scrub that is despised by beast and man, by taxonomist and ranger.

And yet in one area of this zone the tree Psidium cattleianum var. lucidum is moving in to replace these scrub species and in turn acquiring an understory of native forest trees, as economically important a vegetational development as any observed by the author. In many areas the scrambling fern Dicranopteris linearis is especially aggressive following disturbance, blanketing the land and even small trees, so that some have been misled into considering it a climax.

Farther up the slopes occurs Acacia koa (Hawaiian mahogany), which by its growth form, height and the quality of its wood is the only important native timber tree, with a relation to past fires and with silvical characteristics that are not yet well known. The typical rain forest of Hawaii is predominantly composed of Metrosideros collina subsp. polymorpha, a tree which generally does not exceed 50 feet in height and 12 inches in diameter. Associated with it are a large number of endemic species, several of which have excellent potentialities, at least for watershed cover. On the gentle slopes of the larger islands, this same zone supports a range industry which is third only to sugar and pineapple in the islands' economy.

ABOVE THE CLOUDS

Vegetation exists above the clouds in the Solomons, Hawaii and New Guinea and perhaps in other islands, but its organization into plant communities and zones, its prehuman past and its possible future and the location of an effective frost line are all so imperfectly known that few generalizations can be made about them. In Hawaii frosts are not infrequent as low as 4,000 feet and ice formation is appreciable at 7,000 feet.

Much of alpine New Guinea, at least to 12,000 feet, is open and treeless, convincingly interpreted by Lane-Poole 25 as originally forest, but now fire-caused grassland. Here native agriculture is not the chief offender, but fire used as a tool in hunting the wallaby. The higher elevations in Netherlands New Guinea have other zones that are still unknown.

In the Hawaiian Islands the picture is again confused, for here feral grazing mammals have altered the vegetation to an unknown degree and it is doubtful if there are any truly alpine forest species. This paucity of the flora veils the present position of the frost line in the native vegetation. Several thousand feet of elevation are involved, in which Sophora chrysophylla, an important source of fence posts, is the most abundant woody plant. These high regions have probably some of the most intriguing forestry potentialities in all of Oceania.

Forest Influences

While forest influences have long been a matter of abundant fact and fancy, their significance in tropical islands is an unexploited field of research. The fundamental thesis is that the vegetation in the normal course of its development produces certain changes in the microsite and the local environment -- for example, it permits soil accumulation and increasing water infiltration and retention, which in turn, enable the vegetation itself to develop still farther.

Because of water transpired, of absorption and reflection of radiant energy from the foliage canopy and of the resultant effects on the temperature, moisture and saturation deficits of the air above, possible alterations in the regional rainfall are also involved.

Animals introduced on small islands may destroy the vegetation, then die themselves, leaving a sandy barren. Rainstorms coming across the ocean may evaporate in the hot air over an island, but yield precipitation on each side. Clouds passing across the mountains may persist over forested coves and forest plantations, but evaporate over anthropically induced savannas and grasslands.

Drizzles may dry up over a roadway, as reported from Martinique in the West Indies, but fall on the forest on either side. Even a nontechnical observer will be astonished in the tropics at the contrast between a shady, dank, cool, deep- and moistsoiled forest and the scorching rocky man-made desert adjacent.

In short, the magnitude of what vegetation can do to bare land on a tropical island is on a scale for which experience confined to temperate regions does not prepare us. The remedial potentialities of revegetation in low latitudes have been only dimly perceived in research and in practice.

FOREST MANAGEMENT AND USE

UTILIZATION

The original uses of the forest were generally unorganized and unplanned. The natives knew their trees and chose carefully the species and individuals best suited for house posts, canoes, paddles, fishnet floats and other articles to meet their daily needs. The breadfruit was recognized as one of the most valuable of timber trees (propagated vegetatively and existing in endless clonal strains), but was seldom cut, because its fruits were used for food. Coconut also served numerous purposes. Other trees of the strand were selected and many species in the coastalplain forests of the high islands were discriminately chosen, although reports concerning the same species from different island regions vary markedly.

Before World War II organized forestry and forest exploitation can hardly be said to have existed in Oceania. There were a few sawmills on the larger islands, but they served immediate local needs only.

During the war exploitation was stepped up to a high pace to serve the armed forces, especially in the larger Melanesian islands. As in so many tropical regions, the number of species here is great and bewilderingly complex and many of the species have native names only and are unknown to scientific classification.

The wartime operations helped to advance our knowledge of the dendrology and physical characteristics of these forests farther than throughout all previous history, but unfortunately most of such information exists only in unpublished reports and in the minds of the foresters who were directly involved.

RESEARCH

The forestry and agricultural organizations are carrying forward only limited progrmmes of basic forestry research. Usually these organizations are under pressure to serve immediate needs and hence their investigations are oriented toward the planting of exotic tree species of proven worth elsewhere and toward the control of actively destructive factors, such as feral mammals.

On the other hand, scientific research in Oceania is beginning to advance rapidly along general lines and this should benefit forestry in the long run.

Forestry in Oceania involves the management of lands originally or potentially covered with trees and such products of forest land as water and wildlife may locally be more important than commercial timber. For these reasons a sound forestry programme must go beyond the range of conventional silvicultural activities.

It must include interpretations of the vegetation of pre-European and pre-native times and also serious studies of the remaining samples of natural and semi-natural vegetation, which in some instances have very desirable characteristics.

The artificially induced forest types, such as those from plantings, should also be investigated with regard both to their continuing or self-perpetuating characteristics and to their values in other respects than for timber production. Because of the extraordinary complexity of the native and introduced floras and because of the wide range of distinctive environmental conditions within relatively small areas, forestry in Oceania offers unsurpassed challenges to the ingenuity of the practical researcher.

VALUE OF THE FORESTS

Timber holds first place among the forest products of New Guinea and Fiji and may do so in the future in the other continental islands, as is generally true on the continents. On the oceanic volcanic islands, however, the native forests bear very few species or individual trees of such size and quality as are suitable for commercial timber and experiments are constantly progressing with introduced plants. The aborigines use wood for house posts, paddles, canoes and sundry minor articles, for which individual trees are selected when needed.

In Hawaii the chief value of the Forest Reserves is in relation to watersheds, because of the demands of the sugar and pineapple plantations and of urban communities and the same probably also holds true for the other high islands under more intensive use.

Where there is mangrove vegetation, it is extensively exploited for fuel and in Fiji is under a 40-year management cycle, as we have seen.

On most of the islands the main use of secondary growth is for firewood and charcoal. For fence posts, highly important in all grazing regions, the naturalized Prosopis chilensis is one of the chief sources of supply. There is no pulp industry of importance, although recent developments in the use of bagasse are significant. Rubber has been tried in the southwestern islands, but not successfully. Minor forest products are of no more than local importance. For the New Guinea native the principal value of the forest lands is as agricultural soil and once the grasses take over, he moves on, or vanishes.

OWNERSHIP AND ADMINISTRATION

Organized forestry, private or governmental, exists in only a few of the island groups. During the Japanese administration of Micronesia, forestry activities were accelerated, with local sawmill operations, minor exports to Japan and a certain amount of reforestation, especially with Casuarina equisetifolia. The present forestry situation in American Samoa and Micronesia has recently been surveyed.

New Caledonia, Fiji (where a Forest Department was established in 1938) and Hawaii have, or have had, some government administration, with Hawaii far in the lead.

In Hawaii a Board of Commissioners of Agriculture and Forestry has existed since 1903. Its activities have centered largely around the establishment of reserves almost entirely for watersheds, the elimination of the hordes of feral mammals, fencing, a very elaborate planting programme that has involved a vast number of alien forest species and the controversial practice of introducing game animals. In the early days, forest plantings were sometimes made in total disregard of basic silvical principles and failures were excessive. For a long time, success was gauged on survival and rapid growth and undoubtedly this is a chief factor in the case of a timber species that is to be harvested. For self-perpetuating forests, however, this policy may be very deceptive. An intensive report is much to be desired on all these forest plantings of the past, not only with regard to the survival of planted individuals, but also as to whether different species are capable of spreading naturally. Some species may be found to act as "nurse crops" that is, to prepare the way for the introduction of another semi-natural plant community closer to the original self-perpetuating forest.

DESTRUCTIVE FACTORS AND SHIFTING CULTIVATION

The activities of the shifting cultivator have contributed much to the establishment of certain of the present plant communities. Owing to its good accumulation of humus adapted to crops, high forest is invariably cleared and burned and thousands of cubic feet of valuable timber go up in smoke every year. This method of cultivation, which ignores sound agricultural principles, is responsible for covering large areas with secondary forest (jungle) of no commercial value, or else with grasses. Accelerated soil erosion in such areas is enormous, especially on steep slopes.

Tradition and lack of suitable land to satisfy all chena (shifting) cultivators have prevented the government from taking drastic steps. To localize shifting cultivation, however and thereby to prevent the wholesale devastation of valuable forests, the government has divided the chena land into blocks, each of which is cultivated once in ten or twelve years and has also made available irrigable land for permanent cultivation. Areas subject to heavy chenaing are now being reforested on a co-operative basis according to the "taungya forest system," which was first developed in Burma in 1856 and has been adopted in Java and other countries of Southeast Asia. Under this system government land is leased out rent-free for the raising of short-term food crops and in return the lessee raises approved forest species along with his crops.

COMMERCIAL PLANTATIONS

Clearing for commercial plantations has been the second most important destructive human factor. Since 1873 more than half a million acres have been opened up for tea plantations and since 1876 nearly 700,000 for rubber (Hevea brasiliensis). Most of these areas were previously covered by forest. Soil erosion on the plantations is heavy and has adversely affected rivers, tanks, irrigation channels and the plantations themselves.

ILLICIT FELLING

A third factor altering the composition of plant communities in most forest areas is illicit felling and the extraction of poles for agricultural and building purposes. Illicit felling is locally looked upon as an honest trade, one that has come down from father to son. Extraction of poles, seldom properly supervised, tends to destroy nearly all valuable timber species in the round-pole stage. These sizes are extensively used for fencing cultivated land and the construction of semipermanent buildings. Despite the fact that "the round pole of today is a log of tomorrow," removal of poles in other than silvicultural operations is still permitted.

Forest History and Administration

When Ceylon was ruled by the Sinhalese kings the entire forest estates of the island belonged to the Crown and large areas were cleared for agriculture based on irrigation. After the eighth century, however, the forest came back, owing to the destruction and neglect accompanying continuous wars with invaders from the continent and to the ensuing ravages of malaria. The forest at first reclaimed the Anuradhapura area, including the large city of the same name and, a few centuries later, the city of Polonnaruwa.

But in the later forest history of Ceylon "man the destroyer" features prominently. Export of timber began with the Portuguese occupation in the sixteenth century.

The Dutch destroyed much forest for the cultivation of cinnamon, in which they carried on a lucrative trade and after 1830 the British undertook timber exploitation on an enormous scale. After several million cubic feet of the best timbers had been exported, a forest policy was initiated under British rule in 1885 and since then conservation and propagation of the forests have been promoted by the government.

The total forest cover is approximately 70 per cent of the island's area. Forest administration is at present the responsibility of four different bodies. The major portion of the forests is administered by the Forest Department, with a Conservator as head. The Department of Wild Life administers forests in the national reserves and sanctuaries for the flora and fauna. The Gal Oya Development Board, an autonomous body created for the multipurpose development of the Gal Oya valley, administers the forests in that area and the Revenue Department administers village forests set apart for domestic needs. Private forests form a negligible part of the total forested area.

Care of the Forests

Timber extraction for domestic and commercial purposes has been heavy in the past, but the inroads made during and since World War II have contributed most toward the present depletion of timber stocks. A considerable period of rest and cultural treatment is necessary for the majority of the forests, except where inaccessibility, lack of drinking water in the dry seasons and labour difficulties have prevented exploitation and thus helped to protect the stock. In the tropical evergreen zone the forests of the Sinharaja region are still intact. The stocks here, which consist mostly of Dipterocarpus zeylanicus (hora), could, on a rough estimate, produce half a million cubic feet yearly. In the dry-forest zone, the main source of valuable timbers, the only unworked forest is one of about half a million acres in the Northern Division (northwest of Trincomalee).

SILVICULTURAL OPERATIONS

Creepers and unwanted trees are removed by felling, girdling, or poisoning and thinning is confined to plantations. Plantations in the dry and intermediate zones were started in 890, as part of a plan for the co-operative reforestation of areas opened for shifting cultivation. About 15,000 acres of good forest have been planted. Species introduced with varying success are Artocarpus integrifolia (Oak), Swietenia macrophylla (mahogany), Tectona grandis (teak), Azadirachta indica (margosa), Vitex pinnata (milla), Felicium decipiens (pihimbiya), Alstonia macrophylla (havari-nuga) and Berrya cordifolia (halmilla).

In some areas the forest stock has been improved by planting Berrya cordifolia in small clearings scattered within the original forest. These operations have been fairly successful. The afforestation of the grasslands (patanas) of the central hills comprising the montane zone was begun in 1929 and approximately 5,000 acres have been successfully planted with exotics (Australian gums and some conifers). Although no previous tests were made of the suitability of the species to local conditions, the choice has mostly proved good. The species that have shown themselves most adaptable are Eucalyptus robusta, E. microcorys F. Muell. and E. saligna Smith.

After trials the planting of E. globulus,E. acmenioides Schauer and E. maculata was abandoned. Some areas in the montane zone have been planted with conifers, mostly Cupressus macrocarpa and trials have been carried out with Juniperus procera Hochst ex. Endl., Araucaria and Pinus spp. Acacia melanoxylon R. Br. has been successfully raised, but only on a small scale. Conifer planting was abandoned for some years owing to the extreme susceptibility of the plantations to damage by fire. A fresh experimental attempt, however, is being made with Pinus caribaea Morelet. Except where strong winds prevail, all plantations on up-country grasslands show good promise. With the cooperation of the FAO, the Forest Department has embarked on an extensive scheme of mechanized afforestation of the patanas and low forests and at present the target is a minimum of 2,000 acres per year.

Chief Timber Uses

In Ceylon timber is used not only for building construction, which makes heavy demands, but also in the manufacture of plywood, furniture, chests, safety matches, toys and railway sleepers. The demand for nondurable light hardwoods is on the increase. A plywood factory run by the government utilizes such species as Doona congestiflora (kokun), Calophyllum bracteatum (walu-kina), Palaquium petiolare (kirihembiliya) and Myristica dactyloides (malaboda). More than 150,000 cubic feet were used for this purpose in 1952. Species consumed in the manufacture of chests, toys and matches are all nondurable varieties such as Alstonia scholaris (ruk-attana), Canarium zeylanicum (kekuna) and Antiaris toxicaria (riti).

The furniture industry, on the other hand, uses the most valuable durable timbers and there is a yearly demand for approximately 200,000 sleepers (600,000 cubic feet), which can be made only from primary hardwoods. Wood is also extensively used as fuel in the processing of tea and rubber and for domestic purposes, but this requirement should eventually be met by the development of hydroelectric schemes, one of which (Laxapana) is already under way.

The timber position of the island as a whole is unsatisfactory, extraction at present exceeding the yearly increment. The population of Ceylon, now well over 7,000,000, is growing rapidly and at the present rate of increase will double itself in another 25 years. Concurrently the demand per capita for timber, now approximately 10 cubic feet, is also rising.

The problem of future supplies cannot be solved solely by afforestation and better forest management, which can never keep pace with the demand. During the last dozen years there has been an alarming increase in the import of timber, the value of which has jumped from Rs. 193,000 in 1942 to Rs. 21,000,000 in 1952. The only solution is the utilization of species that have hitherto been treated merely as forest weeds. The Gal Oya Development Board has taken the first step in this direction by installing a plant for the pressure treatment of timber and even secondary species (light hardwoods) are now used successfully for building purposes. The Forest Department has also started experiments in timber treatment in a small experimental pressure plant and in open tanks and steps are being taken to install a plant similar to that of the Gal Oya Development Board.

With a view to the utilization of waste timber material from sawmills and carpentry workshops and of branch wood and inferior species, efforts are now being made to develop building bricks from a mixture containing timber and cement and production on a small scale has already been begun.

Governmental and Public Interest

Of recent years both government and the public have shown an increasing awareness of the value of forests as a national asset and of the important role they play in the country's economy. To raise the standards of forestry a forestry field-training school was established in 1951 with two training centers, one in the wet and the other in the dry zone and considerable research work is carried on by the Forest Department. The Soil Conservation Act, passed in 1951, empowers the Government to enforce anti-soil-erosion measures even on private lands. All these steps augur well and it is expected that in the not too distant future the present unsatisfactory timber position will be remedied.

There are three national parks, two of which-Ruhunu and Wilpattuhave been in existence since 1937. These abound in the fauna of the island and provide excellent relaxation and recreation for young and old. Ruhunu in particular, which is only 176 miles from the port of Colombo, has earned popularity in many parts of the world and even casual visitors to the island find time to spend a few hours here. The third and most picturesque park, recently created under the Gal Oya scheme, shows promise of great popularity. The island has some of the most beautiful scenery in the world and forest work is made especially interesting when the forester finds deep in the forest, as he often does, ancient ruined temples and other monuments to a splendid past.

EXTRACTION AND REGENERATION

In Ceylon today modern machinery and methods applied in biblical times are found side by side. Owing to inadequate staff and consequent lack of supervision, timber contractors used to apply the "cut what you like" policy and the present generation is paying heavily for the sins of its forbears. Now, however, the extraction of timber is usually done on a selective basis and precautions are taken to leave sufficient trees of valuable species as seed bearers.

The heavy intermixture of species renders the removal of valuable timber, which seldom represents more than 5 to 10 per cent of the stock, both difficult and costly. The best forest tractor, one that can not only move easily between bushes and unwanted trees but also bring logs to the roadside and load them into trucks, is still the elephant and elephants are also employed in almost every sawmill yard.

Regeneration of valuable species after extraction is quite satisfactory, although species of low commercial value, such as Xylopia parvifolia (netaw), Ailanthus triphysa (wal-bilin) and Canarium zeylanicum (kekuna), in the wet zone and Ficus spp. and Hemicyclia sepiaria, in the dry zone, tend to overdevelop in areas that have been heavily exploited. Clear-cutting or excessive opening usually results in areas being overrun by forest weeds such as Macaranga peltata (kenda) and Trema orientalis (gedumba).

OCEANIA

Ten thousand islands -- thirty million square miles of sea -- less than 2 per cent of the surface land. One island alone, New Guinea, has three times the land area of all the others. Those others are as if an area equal to that of Great Britain had been fragmented and scattered over one third the world. Such is Oceania. Yet its islands and their forests, specked on some of the earth's most critical airways and seaways, of strategic importance in peace and war, command attention. The origins and relationships of the floras of Oceania have challenged the ingenuity, imagination and fact-finding capacities of botanists.

A drop of only 60 feet from present sea level would join New Guinea and Australia. Such a united continent of Papualand is believed to have existed 30,000 years or less ago. As a result of this or earlier unions, the forest flora of New Guinea bears a marked Australian character, especially as to casuarina, eucalyptus and oak. A lowering of only 60 or 70 feet more would expose the Sunda Shelf, uniting Southeast Asia with Sumatra, Java and Borneo. Such a union in the past must have brought the continental flora far east before its overseas migrations were launched.

THE EFFECTS OF MAN

Black Men and Brown Men. The first peoples to be pushed out of Asia into the islands, either by deterioration of climate during the last glacial period or by more aggressive tribes, or both, were Negroids, Ainoids and Veddoids. As they entered Melanesia ("black islands"), the black fuzzy-haired Negroid element was predominant. Hunters and food-gatherers of an Old Stone Age culture, they probably had no marked effect on the forests of their islands.

Four to five thousand years ago, the New Stone Age culture spread to the Negroid islands of Oceania, a culture that turned the people into farmers practicing a destructive shifting cultivation that has persisted to this day and is of great significance from the standpoint of world forestry. To these folk the forest was and still is a hindrance. They cut and burn the trees and plant crops for a few years, until soil fertility presumably lessens. Then the site is abandoned and a new forest is cut. In the meantime the old site grows up to secondary forest. In a decade this is recleared and reused, but for a shorter period. Fire is omnipresent. Rhizomatous grasses invade, which successive fires only encourage by burning the tree seedlings.

The grassland, encroaching on the surrounding forest even without agriculture, is unmanageable and the farmer moves away permanently to new forest sites. New Guinea today is a patchwork of forest and grassland wherever (which is almost everywhere) a "burn-dry season" occurs, from sea level to the frost regions above the mossy forest. Forest remains only in swamps, along stream sides and where the natives have not yet entered, as in a no-man's land between warring tribes.

When such tribes annihilate each other, the forest replaces the grassland, to be destroyed by some future tribe. Since the roving agriculture is a one-way process, especially with increasing population, this destruction of much timber of high commercial value is a problem of first importance to forestry. In addition, it poses a more general problem: the Melanesian in effect destroys his own habitat, for the grassland is useless to him.

After the New Stone Age had spread to Melanesia, but still before the Christian Era, paler, more Mongoloid people invaded the islands to the north, the area now known as Micronesia ("tiny islands"). And then the far-flung islands from Hawaii to Samoa and New Zealand were populated by the handsome race known as Polynesians ("of many islands"), who conquered vast distances in frail canoes. Ethnologists now consider these folk to have migrated through Melanesia toward the east and to have emerged with their present bronze skin.

The theory that they came from Peru, by the Kon-Tiki route is not fashionable. All these invasions were barely completed by A.D. 1400, the date of the first entering of Hawaii.

The Micronesians and Polynesians radically altered the forests of the high islands with regard to both flora and vegetation. They brought with them a limited number of plants; these included a few weeds, but for the most part were food plants, of which the coconut has done the most to change the landscape. It is now dominant on practically every low island and at the lowest elevations of the high islands too. What the forest was originally, or would be now if left to itself or planted to other trees, is a matter of conjecture and future experimentation.

A shifting upland cultivation, such as that of Melanesia, was not part of the culture on the high volcanic islands and consequently the mountain forests were but seldom entered. On the other hand, fire -- as always with primitive peoples -- was uncontrolled and often started for pleasure and hunting. It is the author's opinion that on many islands fires burned as frequently and as far up the hills as the vegetation and the climate would permit, destroying not only large areas of original forest but also the evidence of present forest potentialities.

White Men and Yellow Men. In 1520 Magellan passed through the straits that bear his name and, the following year, landed at the Mariannas. Thus was opened a new chapter in the exploitation of Oceania, a chapter of conflict that continues today between two fundamentally opposed philosophies of social living, a conflict that every forester faces squarely when he tries to administer tribal forest lands.

The original natives lived in small social units. Theirs was an agricultural and subsistence-barter economy, based on few material wants. With the crudest of tools, they fashioned a satisfying existence on infertile specks of land. Living under a form of "communalism," the individual had no incentive to amass personal wealth, for he would only have to share it with his tribe. To this day and for these reasons it is difficult to lure natives to forest work, unless they make a total break with their home villages.

In striking contrast, the new invaders were members of large social units. Theirs was a complex money economy, based on many and evergrowing material "necessities" Under the guise of "individual liberty," each man had every incentive for personal gain and thus wave after wave of Occidentals has passed over Oceania, each intent on its own brand of personal gain: souls, morals and modesty; loot and trade. The invaders streamed in from Europe, the United States and Japan and, when native muscle was not co-operative,

Hindus, Chinese, Malayans, Filipinos, Javanese, Tonkinese and Puerto Ricans were herded in. World War II, however, left Oceania with a new political geography and under a "trusteeship" philosophy that has shown itself more altruistic and more humanitarian than history has usually known. Such elements make the Oceania of today the unusual stage that the forester must know if he is to act effectively.

Sandalwood, 1790-1850, is a finished case history of forest "mining" Undoubtedly, the value of island sandalwood was discovered independently on several occasions and the period of its exploitation is filled with double-crossings. Many a chief succumbed to the money lure, assuming island monopolies and forcing his people to work at gathering the wood in the unhealthy rain forests. Either by nature or with the connivance of the gatherers, who are said to have pulled up all the seedlings they saw, commercial sandalwood disappeared. Apparently the plant has some silvicultural peculiarities, for it has not reappeared in any quantity, either naturally or by artificial means.

The introduction of large naturalized mammals is another aspect of the Occidental invasion that is having a continuous and most important effect on the forests. Only the continental islands possessed a native fauna of large animals; the oceanic islands apparently developed their biotic communities in the complete absence of hoofed animals. It appears that the native forests have extremely shallow root systems, critically located in a moist surface layer. The slightest disturbance to the ground vegetation, as by grazing or browsing, kills the trees, in turn opening the adjacent forest to drying and destroying it, with an effect almost of a chain reaction. Cases are on record in Hawaii where grazing stock was put into a forest for a matter of a few weeks and then removed, but the forest died thereafter. If such destruction were on land planned for grazing, no unexpected problem would result.

Unfortunately the early white settlers introduced cattle, sheep, goats and horses. The islanders placed a taboo upon them, but some escaped to the hills and showed remarkable fertility. In addition, missionaries and traders often released animals on small uninhabited islands, for the sustenance of shipwrecked individuals. The destruction of the native vegetation by these feral animals and by pigs, earlier naturalized, has been tremendous. Wholesale slaughter of thousands of animals has since taken place in some islands; and to this day one of the most useful activities of a local forest service is the building and upkeep of border fences, not only to keep out adjacent high-quality stock, but to prevent reinvasion by feral animals.

Although rabbits have in general been less of a problem, Christophersen has made an interesting report on their history in Laysan, in the Hawaiian Islands. Until 1903 this little island had a flora of 28 species and a solid vegetation of grass and shrubs. In 1911 rabbits were swarming "over the island by thousands" By 1923 only four species of plants could be found and Laysan was described as a "desert of sand," with no mention of rabbits, alive or dead-another example of an introduced mammal annihilating itself by destroying its own habitat.

Much more than half of the flora of some of the islands is composed of species that have been introduced and naturalized since the white invasion. Moist uplands have been less affected, but arid lowlands -- with their original vegetation already destroyed-were fair prizes for hordes of weeds. In many islands the bulk of the lowland vegetation is now composed of alien species and one must search carefully for the indigenes.

Furthermore, on destruction of the native forest by grazing, the land is taken over almost at once by alien grasses and shrubs which, as on the grasslands of New Guinea, form a strong deterrent to the return of the native trees. This situation has given rise to egregious comparisons between the native flora and the native cultures that disintegrate before the onslaught of aggressive aliens. The contrary, however, would seem to prevail in the case of the flora. The writer (1942), in accord with Allen's interpretation for New Zealand (1936), has presented field evidence in support of the thesis that most of the alien plants are pioneer species expanding into "open" lands and forests that have already been disturbed, whereas the native trees are largely of end-stage (climax) species that will hold their own against most of the invaders. The critical need in reforestation is for a "nurse tree," native or alien, which can naturally invade the grassland and scrub, forming a growth that can in turn be invaded by the native forest trees.

Cimates

Oceania comprises three main kinds of islands:

- "Oceanic coral low islands," composed entirely of coral limestone and presumably capping submerged volcanic deposits;
- "Oceanic volcanic high islands," composed of active or extinct volcanoes, arising from the floor of the ocean basin and usually complicated by fringing reefs;
- "Continental high islands," composed of sedimentary strata overlying a granite base, usually complicated by volcanoes and coral reefs.

The low islands partake of what might be called regional climates, controlled by the warm moist trade winds blowing diagonally from the horse latitudes (at about 30° N and S) westward toward the equator. These winds, as they pass over warm, low, sandy islands, may be sufficiently heated so that no rain falls, even when there is precipitation on the surrounding ocean. In general, within five degrees of the equator there is a dry zone, which includes the Phoenix and Gilbert islands and at higher latitudes a moister zone.

At still higher latitudes, the seasonal shifting of the trade-wind belt with the prevailing westerlies modifies the climate, as in Hawaii. In the western Pacific, especially in New Guinea, the influence is predominantly that of monsoons, related to the greater seasonal warming and cooling of large land masses than of the adjacent ocean.

The climates of high islands may vary markedly even within distances of less than a kilometre. The moist trade winds, striking a volcanic island, are deflected upward, where they cool to condensation. Passing over and beyond the summit, they drop, are warmed and dry up. The result is a series of concentrically arranged precipitation belts, placed askew on the contour pattern.

The centre of highest precipitation is slightly to windward of the peak and each rainfall belt is tipped down to windward so that at any one elevation there is less rainfall and a drier vegetation on the lee than on the windward side of the island. Each valley and ridge adds its minor complications. The majority of these climates exhibit a great variety of physiognomic vegetation types, depending largely on human activities, especially management practices. Consequently, any climatic classification which invokes a single vegetation type as the "climax" for its zone runs immediately into inconsistencies.

Soils

Few generalizations can yet be drawn on the soils of Oceania. Those of the continental islands are probably as varied and as complex as in any continental area, though they are today little known.

The low limestone islands have invariably a thin and poor soil, occasionally favored by a layer of organic matter from the vegetation or from bird colonies. More has been published on the soils of the high volcanic islands, especially as regards agricultural Hawaii. It is probable that many of the forest soils are ferruginous laterites. Aside from a surface crust, sometimes present, aluminum and silicon increase downward, iron concentrates at an intermediate position and titanium remains abundant at the surface.

Soil erosion can be critical, often spectacular, but as yet has failed to arouse the governmental authorities, as in other parts of the world. Perhaps this is because most of the erosion has been in unpopulated areas from the start and has not resulted in the displacement of contemporary populations and agricultures.

Much of it apparently was consummated before the advent of the white men, or else immediately after-owing to the naturalization of grazing mammals -- but before European land managers began to look more critically at the land. Within a few miles of Honolulu there was a valley forest a few years ago half buried with debris from the hills. Elsewhere entire coastal plains are said to have been formed in recent years and relict soil patches hanging on rocky mountains attest to past conditions. Unlike that in temperate regions, the decomposition of the volcanic rocks in the tropics may be rapid and some lands may be amenable to management practices that would restore seminatural forests to areas now completely barren. The research on this remains to be done.

Island Types

However, that these are not only incomplete, but are not indicative of potential forest land, since there is much grassland and savanna that could be converted to forest under rational management practices. Such grasslands include not only those of New Guinea and the Solomons, but also those of Hawaii, Guam and even Yapy and Palau.

THE COMING OF THE PLANTS

Although the last word has by no means been said, it would appear that the lesser Pacific islands obtained their original plants from across the seas, with all the hazards and uncertainties and failures of a man seeking to cross in a rowboat and trying to perpetuate himself if he could land. Hurricanes, rafts of floating logs, stray birds and other "accidents," continuing to this day, have contributed to the origins of the present floras.

There is now a consensus that the floristic invasion of Oceania has come from three main regions: the Americas, most obviously in Hawaii; a now vanished Antarctic continent, with the remnants of its flora found in southern South America, South Africa, Australia and New Zealand; and Indo-Malaysia, which in turn is basically Asiatic.

Despite the prevailing easterly winds, however, the tropical cyclones from the west have been most influential. As one proceeds from the western borders of the Pacific, the flora becomes more and more impoverished in genera and appears obviously to be a fraying-out, an attenuation of the Asiatic plant world, as the chances of successful immigration become less and less over the tremendous distances of water.

Fosberg has analyzed the Hawaiian flora with regard to its origins, basing his discussion on the number of assumed original immigrants. For the seed plants, he finds that 40.1 per cent are Indo-Malaysian, 18.3 per cent American, 16.5 per cent Antarctic and the remainder boreal, cosmopolitan, or obscure.Offsetting in part this paucity in genera and larger categories, there has been an extraordinary speciation in certain groups.

Whatever be the causes, the original immigrants, including trees, have sometimes indulged in endless morphologic variation, often with each valley developing its recognizable form.

Certainly many of the named forms are ecologic modifications and many more would interbreed and lose their identity if placed in a related population; but many others are undoubtedly established genetically and the recognition of them as desirable strains in forest trees is an open field for silvicultural research. Immigration and speciation are probably continuing now as in the past, but on too slow a time scale to be of practical importance. The striking effects of introductions by man will be considered later. The above discussion holds for what one might term the "inland flora," for species that cannot tolerate salt, die when submerged in salt water and shun the strand. It is these species which pose the problem of overseas migration to the botanists.

The "strand flora," entirely to the contrary, seems to have evolved capable of overseas migration. Buoyant seeds that withstand many days of floating -- even as germinated seedlings -- are found in totally unrelated plant families. This flora is remarkably constant in all the islands -- evidence of the success of these adaptations; and this flora is adapted not only to migration but also to growth under saline conditions, either in soil or air. Indeed, the plants are frequently incapable of terrestrial migration (though many grow well where planted). For all these reasons, the strand flora forms the entire vegetation on countless low islands. On the high islands it tends to predominate only in the plant communities near the seacoast.

Among all these native plants, as well as among naturalized species, there are innumerable kinds of trees of distinct value to the forester, especially in the islands close to Southeast Asia. No manual of dendrology exists for this territory, although Kraemer has published a volume for part of Malaysia, including some of Melanesia. The book describes more than 100 important species, said to be about one per cent of the total tree flora.

OCEANIC LOW ISLANDS

Where volcanic or other materials have formed the sea floor at depths of less than 150 feet in tropical waters, they have been subject to colonization by coral and related organisms. These have built up calcareous deposits as living reefs at, or slightly below, sea level. Surmounting such a reef, generally at its margins, an atoll is usually a more or less circular group of small islands, produced by wave and wind action, enclosing a "lagoon" To the geomorphologist atoll islands seem ephemeral, subject to rapid destruction, especially during storms. Forestry practice must reckon with the hazards to which such islands are subject.

Table. Oceania: Land Area and Forest Area for Certain Islands, 1950-1953 (In Thousand Hectares)

Land Area	*Forest Area Accessible*	*Inaccessible*	*Total*	
British Solomon Islands, 1953	2,890	20	2640	2,660
Fiji, 1953	1,828	140	815	955
French Oceania, 1950	360	50	65	115
Hawaii, 1950	1,665	450	325	775
New Caledonia, 1952	1,582	10	70	80
New Guinea (Australian) and Papua, 1951	47,400	3,00	30,000	33,000

As a habitat for plants, atoll islands, despite their almost infinite number, their human populations and their emotional appeal, do not hold much for the forester interested in quantity. Their vegetation normally comprises a series of belts from sea to lagoon. Back of the sea beach is a scrub 2 to 5 meters high. Next is a narrow forest zone of Messerschmidia argentea and Pandanus spp. In the Marshalls and Carolines Terminalia litoralis is found in this Messerschmidia zone.

The greater part of the typical atoll is a forest which today is almost entirely composed of Cocos nucifera, topped sometimes by huge breadfruit trees (Artocarpus altilis) where the rainfall is heavier. The timber of breadfruit is excellent in almost every respect, but the tree is seldom cut since it serves primarily for food. It would appear that the original forest vegetation was composed of such trees as Pisonia grandis, Pandanus spp., Ochrosia parviflora, Pipturus argenteus, Hibiscus tiliaceus, Messerschmidia argentea, CaloPhyllum inophyllum, Barringtonia asiatica, Eugenia spp. and others. Pisonia may be the sole dominant on dry islands, such as Vostok. Some of the isolated eastern atolls may have floras composed of as few as five species, or even less, seemingly the result of chance migration.

Along the lagoon beach of most islands is a narrow strip of scattered trees such as Cordia subcordata, Hernandia ovigera, Terminalia catappa, Barringtonia asiatica, Thespesia populnea and, in rocky places, Pemphis acidula. Mangrove species, where found at all, occur in small depressions in the centers of the islands.

CONTINENTAL ISLANDS

The continental islands of Oceania include Fiji, New Caledonia, the New Hebrides, the Solomons, the Bismarck Archipelago and New Guinea. In them are found not only all the varieties of forest characteristic of coral low islands and oceanic volcanic islands, but also their own complexities, due to larger floras, more varied rock strata and topographic features and climatic regimes ranging from slight to complete dominance by the monsoons. Nevertheless, the same basic pattern of arid, humid, cloud and super-cloud zones occurs, if not in the orderly arrangement found on a single high volcano in the tradewind belt.

OCEANIC HIGH ISLANDS

The oceanic high islands display endless variations, but basically they consist of volcanoes. They not only occur in chains and groups but also in such total isolation as that of Easter Island. The original formations were pahoehoe and aa lavas, with bombs, ash, cinders, dust and other ejecta. Erosion may have advanced to such a degree as to leave knife-edge ridges and deep valleys (sometimes amphitheater-headed, because of greater precipitation upstream), footed by coastal plains. Ancient coral reefs, developed at previous higher seá levels, may appear along an island's flanks. Soil maturation has progressed variably from the different parent materials, with the development of distinctive profiles. Despite this profusion of types, a pattern emerges of interest to the forester.

This zones, arranged consecutively from sea level to summit. Each zone, except in the driest, coldest and cloudiest sites, has a characteristic assemblage of forest, savanna and grassland, with certain apparent correlations with contributing factors, although implications of causality are still premature. Sunlight attains a maximum at sea level on the lee side of the typical island and decreases upward, with a marked diminution at the beginning of the socalled "cloud zone," a minimum in the mossy forest and then a rapid increase to a second, high-altitude maximum.

Precipitation varies inversely with sunlight, being low at sea level, maximal in the cloud zone and again low at very high elevations.

Cloudiness, although it may occur at any elevation, is of significance only within the cloud zone, where its intensity increases rapidly to a maximum at its upper level. Altitude, perhaps because of the ease of obtaining data regarding it, is mentioned ad extremum by botanists and foresters, but does not seem to exert any direct effects, at least upon the details of vegetation.

The distribution of floras in relation to vegetation zones has been considered only for Oahu, Hawaii. Despite its significance in forest management, both in control of natural stands and in planting progrmmes, foresters have almost totally neglected this subject. Its importance is illustrated by the fact that in any one zone the predominant trees may be either (a) those that are there capable of maximum individual development and would not thrive in an adjacent zone, or (b) those that have been crowded out of an adjacent zone by competition with others and would actually develop more luxuriantly if planted in the adjacent zone with the competition lessened.

As a working hypothesis for Pacific high islands, areas characterized by six different floras:

- The strand flora;
- A desert flora, which may exist on some islands where there is less than 35 inches of precipitation a year (some widely naturalized Opuntias may occur in this area);
- An arid-region flora on islands having roughly less than 50 inches of rain, with a maximum floristic complexity at the higher precipitation levels and attenuating out below to mingle with the desert plants (tremendous numbers of alien plants have been naturalized in this floristic area);
- A humid-region flora occupying a region of rain forests, with high percentages of endemic species and few naturalized aliens (many of the plants pass upward into the cloud zone, where present);
- A cool-tropical cloud-region flora;

- A flora or floras above the frost line. On some islands, above the cloud zone but below the frost line, the humid-region and arid-region floras are repeated, but in reverse order, as on the island of Hawaii.

The significance of an effective frost line on the islands of Oceania remains to be investigated. One might expect, judging from continental areas, that it would form a fairly distinct separation of tropical from nontropical conditions. The situation in New Caledonia and in the New Hebrides, Society, Bismarck and Solomon islands is unknown. In oceanic Hawaii the flora at high altitudes appears to be nothing more than an extremely impoverished attenuation of plants primarily tropical, with almost no temperate American origins-mute testimony of the failure of aerial migration from that direction. Continental New Guinea, on the other hand, has an extensive super-frost-line flora, but its geographical affinities are poorly known. All the super-frost-line regions hold great potentialities for the introduction and naturalization of temperate forest species.

Each vegetation zone possesses a complex of topographic variations, ridges, valleys and plains that make difficult the recognition of zonal differences. In those cases where a so-called rain-forest zone is predominant on coastal plains and a "montane-forest zone" on foothills and slopes, one has reason to wonder whether a climatically significant zonal difference actually exists.

The precise mapping of the boundary between two distinct zones has been attempted only on Oahu. Here the line was found to pass through a wide physiognomic variety of plant communities and to vary in altitudinal range by almost 1,000 feet, bulging seaward on the ridges, where the cooler temperatures may have offset the lower rainfall to permit an equivalent moisture effectiveness.

6

Forest Hydrological Cycle

Rain *(P)* is the main precipitation input to most forests, supplemented by snow at higher altitudes and latitudes and by 'occult' precipitation (fog) in coastal or montane fog belts. A small part of the precipitation reaches the forest floor directly without touching the canopy, the so-called 'free' or 'direct' throughfall. Another (usually small) part travels along the branches and trunks as stemflow *(Sf)*. A substantial portion of the precipitation intercepted by the canopy is evaporated back to the atmosphere during and shortly after the storm (called interception loss, E_i), whereas the remainder reaches the soil surface as crown drip after the storage capacity of the canopy has been filled. Because direct throughfall and crown drip cannot be determined separately in the field, the two are usually taken together and referred to as through-fall *(Tf)*. However, the distinction between the two is useful in modelling the interception process. The sum of through-fall and stemflow is commonly called net precipitation and is usually substantially smaller than amounts of incident precipitation unless there are significant (unmeasured) contributions by fog precipitation:

$P=Tf + Sf + E_i$

and

$E_i = P-(Tf+Sf)$

If the intensity of the total throughfall and stem-flow reaching the forest floor exceeds the infiltration capacity of the soil, the unabsorbed excess runs off as 'Hortonian' or 'infiltration-excess' overland flow (HOF). Due to the generally very high infiltration capacity of the organic-rich topsoil in most forests, this type of flow is rarely observed in undisturbed forest unless there is an unusually dense clayey substrate or an excessive concentration of stemflow. Not all of the water infiltrating into the soil emerges as streamflow.

A large part is taken up by the vegetation and returns to the atmosphere via the process of transpiration (£$_t$). The term 'evapotranspiration' *(ET)* is used to denote the sum of transpiration (evaporation from a dry canopy), interception loss (evaporation from the exposed surfaces of a wet canopy) and evaporation from the litter and soil surface (£$_s$). The latter term is often small, especially in dense forests where little radiation penetrates to the forest floor, humidity is high and the air virtually stagnant. It is essential to distinguish between E_t and E_i because the former is strongly governed by stomatal control and the latter mostly by the aerodynamic properties of the vegetation. Thus:

$$ET = E_i + E_t + E_s$$

The remaining soil moisture drains towards the nearest stream via 'throughflow', the result of downward-moving water meeting an impermeable layer of subsoil or bedrock and being deflected laterally.

Such water drains slowly and steadily, thus accounting for the 'baseflow' of streams. In humid temperate climates, baseflow generally reaches a minimum in the summer period, whereas under seasonal tropical conditions this occurs during the dry season. These low flows are usually referred to as 'minimum flow', 'dry-season flow' or simply 'low flow'. During rainfall, infiltrated water may take one of several routes to the stream channel, depending on soil hydraulic conductivity, slope morphology and the spatial distribution of soil moisture. 'Saturation' overland flow (SOF) is caused by rain falling on an already saturated soil.

This situation typically occurs in hillside hollows or on concave foot-slopes near the stream where throughflow tends to converge and maintain near-saturated conditions. Occasionally, widespread hillside SOF (*i.e.* outside concavities and depressions) has been observed during and after intense rainfall in the tropics in places where an impeding layer is found close to the surface. Rapid through-flow during storms (subsurface stormflow, SSSF) usually consists of a mixture of 'old' (*i.e.* already present before the start of the rain) and 'new' water travelling through 'macropores' and 'pipes'.

As a result of contributions by SOF, SSSF and, in extreme cases, HOF, streamflow usually increases rapidly during rainfall. This increase above base-flow levels is called 'stormflow' or 'quickflow'. The highest discharge observed in association with a precipitation event is referred to as 'peak flow'. Peak discharges may be reached during the rainfall event itself or as late as a few days afterwards, depending on catchment characteristics, soil wetness and the duration, intensity and quantity of the rainfall.

The total volume of water produced as streamflow from a catchment over a given period of time (usually a month, season or year) is called 'water yield'. The interrelated character of the chief components of the hydrological cycle is summarized by the catchment or site water budget equation:

$$P = E_i + E_t + E_s + Q + AS + AG$$

where Q is streamflow or drainage to deeper layers, AS change in soil water storage and AG change in groundwater storage, with the remaining terms as defined previously. All values are expressed in millimetres of water per unit of time (hour, day, week, month or year). Note that AS and AG may assume positive (gain) or negative (loss) values. In view of the seasonal cycle of soil water and groundwater storages in many areas, the values of *AS* and *AG* tend to approach zero on an annual basis. Therefore, the annual water balance is often simplified to:

$P=ET+Q$

FORESTS AND 'OCCULT' PRECIPITATION

Forest-climate interactions in general have been dealt. Therefore the present discussion does not focus on the extent to which a forest cover generates its own rainfall (and over what scale) but rather on the importance of fog and other forms of cloud moisture ('occult' precipitation) as an additional input of water. Wherever fog impacts a forested area, particularly where the fog occurs in the form of an orographic cloud belt or advective coastal fog, additional moisture is intercepted by plant surfaces (or indeed any other obstacle) and precipitation may occur in the form of 'occult' precipitation or 'fog drip', even if no rainfall is recorded on adjacent open ground.

An extreme example has been described by Aravena *et al.* (1989) whose isotope studies have demonstrated that the frequent occurrence of advective sea fogs along the arid coast of northern Chile has given rise to a patchy forest that, in the almost complete absence of ordinary rainfall, thrives primarily on fog. Several studies of fog drip beneath trees along the US Pacific Coast have shown that the amount is directly related to the area and density of the tree profile and to the degree of exposure of the trees to windblown fog.

During times of fog, fog drip at the edge of a forest may be several times the rainfall in the open. However, amounts often decrease sharply towards the forest interior.

Therefore, reports of increased net precipitation beneath vegetation need to be considered carefully in terms of gauge positions with respect to the edge of the forest. At favourably exposed locations the extra inputs stripped from the fog by trees (conifers especially) may be considerable. Examples include 425 mm in 46 rainless summer days below 18-m tall Douglas fir in northern California , 760 mm annually measured in Hawaii below *Araucaria* by Ekern (1964) and 880 mm annually for tall old-growth Douglas fir in Oregon by Harr (1982). Bruijnzeel and Proctor (1995) and Bruijnzeel (2000) recently reviewed the hydrological characteristics of 'cloud forests' found in orographic cloud belts on wet tropical mountains and reported extremely variable net precipitation fractions for these forests (55-130% of incident rainfall).

As such, although tropical montane cloud forests are supposedly exposed to frequent fog , the associated amounts intercepted by the trees are not always sufficient to raise net precipitation totals significantly above the 70-80% commonly recorded for montane rain forests on cloud-free locations. Such variations may be interpreted in terms of exposure to and persistence of fog, as well as to contrasts in canopy epiphyte biomass.

Quantification of the amount of fog intercepted by a forest is difficult, particularly if the fog occurs together with rain. The usual approach is to compare amounts of net precipitation (often throughfall only) beneath the forest with the rainfall measured in the open or to subtract the latter from the catch obtained with some kind of fog gauge placed next to an ordinary rain gauge. A number of different fog gauge designs have been proposed, including wire harps , cylindrical screens , 1m^2 polypropylene screens and louvered metal gauges.

The through-fall method essentially gives an estimate of 'net' fog drip since it includes an unmeasured amount of water lost to evaporation from the wet vegetation. In addition, the result is site specific.

The use of a fog gauge (regardless of the type adopted) also requires a site-specific conversion factor to relate the catch of the gauge to that of the actual forest canopy. Some progress has been made with physical fog deposition models but the data requirements of these models are such that their use at remote forest locations must remain limited to a few well-researched sites.

Therefore there may be some merit in alternative approaches to the quantification of fog contributions, such as the sodium/chloride budget method or the contrasting stable-isotope signatures of rain and fog water.

For example, by comparing the stable isotope compositions of rain, fog, soil moisture and groundwater with those of xylem water in the dominant species of a Californian redwood forest, Dawson (1998) was able to demonstrate the importance of fog to the different plant groups. During a year with average climatic conditions, 25-50% of the water uptake by the tall redwood trees was shown to consist of fog whereas these figures roughly doubled during a drought year. Under-storey herbs were even more dependent on fog water: 19-50% in an average year and 28-100% in a drought year.

Transpiration (£$_t$) ceases whenever the canopy is fully wetted. Because of this and the reduced radiation and high relative humidity normally associated with such conditions, E_t of trees exposed to frequent fog is low. Together with the additional inputs via occult precipitation, this makes for an especially favourable water balance, *i.e.* amounts of stream-flow or groundwater recharge will be comparatively high. For these reasons Zadroga (1981) expressed the fear that clearing tropical montane cloud forest could well result in substantially reduced water yield, especially during the dry season. Some support for this contention could come from the contrasting dry-season flows emanating from two pairs of cloud-forested and cleared catchments in Guatemala and Honduras reported by Brown *et al.* (1996).

However, both catchment pairs were too different in size and elevational range (and therefore exposure to fog and rainfall) to be sure that the approximately 50% reduction in flow was entirely due to the replacement of the forest by vegetable cropping.

A more convincing case was provided by Ingwersen (1985) who observed a (small) decline in summer flows after a 25% patch clearcut operation in the same catchment in the Pacific North-west region of the USA for which Harr (1982) had inferred an annual contribution by fog of about 880mm. The effect disappeared after 5-6 years.

Because forest cutting in the Pacific North-west is normally associated with strong increases in water yield , this anomalous result was attributed to an initial loss of fog stripping upon timber harvesting, followed by a gradual recovery during regrowth.

Interestingly, the effect was less pronounced in an adjacent (but more sheltered) catchment and it could not be excluded that some of the condensation not realized in the more exposed catchment was 'passed on' to the other catchment. The observation of Fallas (1996) of enhanced throughfall in patches of forest surrounded by pasture in the montane cloud forest belt of northern Costa Rica is probably pertinent in this respect. Further (process-based) work is needed to decide on the importance of occult precipitation for water yield from fog-affected areas.

THROUGHFALL, STEMFLOW AND INTERCEPTION LOSS

Amounts and Measurement of Rainfall Interception

Arguably, no subject in (temperate zone) forest hydrology has received as much attention as the measurement of throughfall (*T/*), stemflow *(Sf)* and derived estimates of interception loss (E_i). Yet, until the pioneering work of Rutter and colleagues in the 1960s, the increase in understanding of the underlying processes was not commensurate.

Before then, interception was mainly studied in terms of tree species and age, or the amount, intensity and duration of the rain. For example, in a review covering more than 50 interception studies in the USA spanning half a century, Zinke (1967) arrived at essentially the same generalizations as Kittredge (1948) and Horton (1919) had before him. These can be summarized as follows:

1 E_i is a function of incident rainfall *(P)* and typically declines more or less hyperbolically when expressed as a percentage of *P;*

2 both E_i and the canopy storage capacity *(S)* are generally larger for coniferous forests (E_i = 15-40%; *S* = 0.5-3.0 mm; average ~2mm) than for deciduous forests (growing season, £$_i$ = 20-25%; *S* = 0.5-2.0 mm; average =1.0 mm);

3 winter values of E_i and *S* for deciduous forests are roughly half to two-thirds of summer values;

4 *Sf* generally constitutes only a modest fraction of *P* (typically < 3%), with the exception of smooth-barked species such as beech or maple (5-12%) or young trees;

5 E_i increases with stand density;

6 snowfall interception storage by coniferous canopies exceeds that for rainfall.

The extensive reviews of early European work by Delfs (1955) and Molchanov (1963) and of British studies by Hall and Roberts (1990) broadly support the above generalizations obtained for forests in the USA.

The main contribution by Rutter and coworkers to our current, improved, understanding of the rainfall interception process has been to move away from the empirical site-specific approach of the past to a more explicitly physical description of the process. Central to the Rutter approach is a running canopy (and trunk) water balance in which hourly inputs (rainfall) and outputs (drainage, evaporation) determine the amount of water stored on the canopy (and the tree trunks). The latter governs both the rate of drainage from the wet canopy and the rate of evaporation (for storage less than *S*).

Usually, the results of interception studies are expressed in relation to gross rainfall, either as a ratio, percentage or through various types of regression equations. Comparisons between different studies, even for the same species and age class, are rendered difficult because of the more-or-less unique character of each forest stand in terms of density, presence or absence of undergrowth, exposure to prevailing air streams, and rainfall regimes. In addition, there is a methodological problem in that the notoriously uneven distribution of *Tf* (both in space and time) requires a rigorous sampling strategy that not all studies have achieved. In general, the spatial variability of *Tf* increases with forest density, *i.e.* it is greater in summer than in winter for deciduous stands and much larger in tropical rain forest than in temperate plantations.

The vegetation in most natural forests is made up of a mosaic representing different stages of growth, ranging from young rapidly growing trees in gaps to old-growth emergents on the decline. Burghouts *et al.* (1998) reported an inverse relationship between amounts of *Tf* and litterfall in a spatially heterogeneous lowland rain forest in Borneo. Delfs (1955) observed a gradual decrease in *Tf* over an age sequence from young regenerating spruce in the sapling and pole stages to mature stands.

At the microscale, Eschner (1967) reviewed a range of studies in deciduous and coniferous forests showing that beyond the immediate sphere around the trunks (where *Sf* and drip from the points of juncture of branches and bole tend to concentrate), *Tf* often increases away from the trunks to reach a maximum just within the perimeter of the crown. Hutchinson and Roberts (1981) reported that 98% of all *Sf* generated by a 9-m tall Douglas fir originated in the upper half of the tree, whereas the sheltered and more depressed branches of the lower half (representing 31% of total branch interception area) contributed little to stemflow production. Czarnowski and Olszewski (1970) demonstrated that using up to 30 fixed standard gauges gave different average *Tf* values for a natural deciduous forest in central Europe, whereas the estimate stabilized after using 30 gauges or more. Helvey and Patric (1965) suggested that six gauges during the dormant season and 15 during the growing season would be sufficient for a proper estimation of *Tf* in mixed hardwood stands in the eastern USA, provided the gauges were relocated regularly. Other investigators relied on increasing the sampling surface area, be it through the use of large metal plates , elongated gutters or plastic sheets. However, the use of a large collecting surface *per se* is not a guarantee of a good estimate of E_i. For example, Bruijnzeel and Wiersum (1987) used 10-12 trough-type *Tf* gauges (with a total surface area equivalent to 140-168 standard gauges) in an *Acacia auriculiformis* plantation in Indonesia. Measurements were made during two consecutive rainy seasons with fixed trough arrangements that differed between the two periods because the gauges were removed during the dry season. Although both estimates of seasonal *Tf* had a standard error of less than 5%, the mean values were very different.

Because the difference was only partially explicable in terms of increases in canopy development and differences in rainfall patterns, it had to be concluded that the two fixed trough configurations sampled the spatial variation in *Tf* differently. Such findings indicate the need for a sufficiently large sampling plot commensurate with the heterogeneity of the forest under study.

A 10 x 10 m sampling plot could be sufficient for a uniform tree plantation of high stem density but entirely inappropriate for a heterogeneous species-rich lowland rain forest. Lloyd and Marques (1988), using a 100 x 4m sampling transect in a central Amazonian rain forest, showed that about 40 fixed standard gauges would be needed to sample *Tf* with a standard error of 10% but that the same result could be achieved with only 10 'roving' gauges.

Interestingly, Lloyd and Marques (1988) obtained an E_i of only 9% using 36 roving bottle gauges where Leopoldo *et al.* (1982) had previously reported a value of 19% for the same forest on the basis of 20 fixed collectors. The large difference between these two estimates may well be caused by a more representative sampling of so-called 'drip points' (*i.e.* $Tf > P$) by the roving-gauge arrangement. Not many studies have used a roving-gauge approach and published estimates of E_i must therefore be viewed with caution.

Despite the large variation encountered in the deciduous group, E_i in deciduous stands is invariably lower than in coniferous forest for the same amount of *P*. A similar contrast was noted for coniferous (E_i > 20-25%) and (semi)deciduous plantations such as teak and mahogany in the tropics (£^20%; Bruijnzeel 1997). Although evergreen, the relatively light crowns and smooth stems of (young) eucalypt plantations intercept only modest quantities of rainfall (typically about 10-15%), both in the tropics and in their native Australia.

However, recent work in south-eastern Australia has shown that E_i in an age series of natural eucalypt forest may rise to a peak of 25% at age 30 years after which it declines to about 15% at age 235 years. Although some of the variation in E_i between different deciduous forests, may be related to the presence or absence of a well-developed understorey, little is known of the intercepting capacity of the latter. Noirfalise (1959) obtained values of 10.4 and 12.6% for E_i below bracken fern *(Pteridium aquilinum)* between late May and mid September and between mid September and mid November, respectively. Using rather more indirect means, Gash and Stewart (1977) derived a value of 12.3% of *P* for bracken fern below Scots pine in the eastern UK between mid May and early October, representing 12.6% of the annual E_i of the forest.

Bruijnzeel (1990) derived an average *Tf* of 85% (range 77-93%) from 13 investigations conducted in tropical lowland rain forest that used either a roving-gauge arrangement or more than 20 fixed gauges.

Stemflow in these forests was usually less than 2%. Corresponding average values for tall montane forests experiencing little to no cloud incidence are 75% (range 67-81%) and less than 1%, respectively. Net precipitation in mossy montane cloud forests may be much higher due to the effect of 'occult' precipitation. The most stunted forms of ridge-top montane cloud forest, which frequently have crooked stems and trees with multiple stems, may exhibit exceptionally large amounts of *Sf* (up to 18%; Bruijnzeel 2000).

It should be noted that no distinction was made by Bruijnzeel (1990) between forests at mid-continental vs. continental edge and island locations. Shuttleworth (1989) advanced the idea that forest evaporation (notably E_i) may be much greater in the latter situations compared with the former. He based his contention on a comparison of the micrometeorology of a lowland forest in central Amazonia and a spruce plantation in Wales, UK , as well as on the reported contrast in E_i between the Amazonian forest and a secondary lowland rain forest on the island of Java, Indonesia. The argument was rendered inconclusive at the time because of the use of markedly different methodologies for the estimation of E_i in Indonesia and Amazonia. However, since the late 1980s a number of rainfall interception studies in tropical forests located at continental edge and island locations with high rainfall (>3000mm annually) have confirmed the possibility of very high E_i under such wet 'maritime' tropical conditions, both in the lowlands and at higher elevations.

Litter Interception

Helvey and Patric (1965) argued that evaporation from the litter layer (£$_s$) may constitute a significant component of the overall interception loss and should therefore be determined separately. When measured *in situ* (as opposed to using litter trays which tend to give overestimates), £$_s$ was shown to reach 2-5% of incident *P* in hardwood stands in the eastern USA (=50mm annually). Interestingly, litter drying rates in winter were higher than in summer, presumably because the leafless condition of the forest permitted increased ventilation and irradiation of the forest floor at a time of maximum *Tf*. Later micrometeorological work by Black and Kelliher (1989) in Douglas fir forests confirmed the importance of radiant energy and degree of turbulence to forest floor evaporation.

These authors also showed that $£_s$ and under-storey *ET* are largely complementary processes, with high values of $£_s$ occurring in stands having little to no understorey vegetation and vice versa. Using weighing lysimetry Schaap (1996) showed $£_s$ in a dense but well-ventilated Douglas fir forest in The Netherlands to be mainly dependent on forest floor moisture content, with (modelled) average annual values ranging from 0.21 mm daily on drier microsites to 0.38 mm daily on wetter sites. The corresponding stand-averaged annual total of 112mm represented about 12% and 22% of incident *P* and *Tf*, respectively. Similar values (9.5-11% of *P*, or = 155 mm yearly) were obtained for plantations of *Pinus caribaea* under warm non-equatorial tropical conditions in Fiji. Conversely, much lower annual totals (50-70 mm) have been reported for dense equatorial rain forest.

Interception Modelling

Despite the numerous studies of rainfall interception conducted prior to the 1960s, little progress had been made in understanding the physics underlying the observed contrasts and inconsistencies in E_i, both between and within species and events. Explanations offered in the early literature were usually worded in terms of differences in canopy storage capacity (in turn related to canopy density, deciduousness, etc.) or the intensity and duration of the rain. However, by the mid 1960s profound changes were to occur. In the same year that Helvey and Patric (1965) still suggested that two simple linear *Tf-P* regression equations were sufficient to adequately quantify E_i for hardwoods in the eastern USA under a wide range of conditions, Monteith (1965) advanced a more physical description of the process of evaporation from a vegetation cover while Rutter (1967) presented a process-orientated approach to the study of E_i at the 1965 International Symposium on Forest Hydrology.

The regression approach of Helvey and Patric (1965) was later criticized by Jackson (1971, 1975) for not incorporating rainfall intensity and duration or the length of intervals beween storms. However, Jackson himself was not very successful in raising the coefficients of determination (R^2), and therefore the predictive capacity, of his own more elaborate regression equations, thus illustrating the limitations of the empirical approach. Arguably therefore the two papers by Monteith (1965) and Rutter (1967), complemented by the interception modelling work of Rutter *et al.* (1971, 1975), mark a turn from the previous largely empirical experimental approach to a more physically orientated, process-based, approach. With increasing vegetation height there is a corresponding increase in roughness length *d* and displacement length z_0

and thus, for a given wind speed, a decrease in r_a. Also, r_a decreases with wind speed; approximate values for different vegetation covers at a wind speed of 2.5 m s^{-1} measured at 10m above the surface are 115, 50 and 10-15 sm^{-1} for short grass, field crops and forest, respectively. Because values of the surface resistance to evaporation from a dry canopy (r_s) for grass and forest are much more similar, the magnitude of r_s/ra in Eqn 12.6 is larger for a forest cover. This would lead to lower values of forest *XE* were it not for the compensating effect of the lower albedo and thus R_n of the forest. As a result, as long as soil water is not limiting and atmospheric humidity deficits remain low, transpiration rates for grass and forest do not differ much. This is in great contrast to the finding that as a result of the much lower r_a of the forest, evaporation from a wet forest canopy will proceed much faster than from a wet grassland.

Wet canopy evaporation rates of 0.2-0.5 mm h^{-1} have been observed under temperate conditions, even in cloudy winter weather when XE_{wet} must greatly exceed net radiant energy. Under maritime tropical conditions, values as high as 0.7-1.2mmh^{-1} have been inferred from measurements of net precipitation. As shown by Rutter (1967) and Stewart (1977), temperatures of wet forest canopies may be slightly cooler than the air passing overhead.

This, together with the low aerodynamic resistance conducive to rapid evaporation, allows the development of a downward sensible heat flux capable of maintaining evaporation rates well in excess of available radiant energy. The degree to which the phenomenon influences annual E_i obviously depends on the frequency of wetting of the canopy (*i.e.* rainfall regime) and thus on the overall setting of the forest. For a pine plantation in the relatively dry eastern part of the UK, Stewart (1977) measured instantaneous rates of XE_{wet} of 127% of net radiant energy but on an annual scale E_i was modest at 214 mm or 38% of total forest *ET*.

For a spruce plantation in the more maritime setting of wet central Wales with frequent passage of warm frontal rain, Shuttleworth and Calder (1979) reported instantaneous rates of XE_{wet} of about five times the available radiant energy. Annual *ET* exceeded R_n by 12%. Not surprisingly, E_i made up a large portion (61%) of total forest *ET*. Recently, Schellekens *et al.* (2000) reported an even more extreme example from eastern Puerto Rico, where annual E_i from a colline rain forest varied between 1365 and 1790mm (39-49% of *P*). Although no direct measurements were made of a downward sensible heat flux, there were strong indications that non-radiant energy dominated wet canopy evaporation. Not only were net precipitation fractions not statistically different for daytime and night-time events but also the average evaporation equivalent of net radiant energy (0.11mmh^{-1}) was far smaller than the value of XE_{wet} as inferred from measurements of E_i (0.93-1.13mmh^{-1}). In addition, annual *ET* as estimated from the water budget of a small watertight catchment exceeded R_n by 42%.

The origin of the extra energy is still a matter of speculation. Advected warm air from the nearby Atlantic Ocean is a likely source although, unlike the situation in Wales, wind speeds in eastern Puerto Rico are generally low. Another possible energy source may be heat released upon condensation of water vapour in the air above the forest. Some support for this contention comes from the observation that *Tf/P* was independent of storm size, *i.e.* larger rainfall events were accompanied by higher interception totals. This would suggest a positive feedback of rainfall amount (and thereby condensation) on the magnitude of E_i. However, even if it is accepted that for situations like those encountered in Wales and Puerto Rico, where the total energy available for evaporation during rain can be much larger than provided by radiant energy alone, it is not clear how the aerodynamic resistance of the forest can be so low as to enable these high wet canopy evaporation rates. For the spruce plantation in Wales, Calder (1990) required an optimum value forr_a of 3.5sm^{-1} (vs. a theoretical value of 5.4s m^{-1} using Eqn 12.9 at an average wind speed of 3.75ms^{-1}).

Using Eqn 12.10, Schellekens *et al.* (2000) obtained values of 2.1-3.2 s m^{-1} (depending on VPD) vs. 18-22sm^{-1} when using Eqn 12.9. Whilst in the Puerto Rican case r_a may be lowered somewhat by the irregular character of the canopy due to former hurricane damage, Calder (1990) and McNaughton and Laubach (1998) have pointed to the possibility of enhanced upward transport of evaporated moisture (*i.e.* low r_a) by wind gusts and eddies. Such gusts would assume extra importance under conditions of relatively low wind speeds such as in Puerto Rico.

Determination of A£$_{wet}$ using eddy correlation equipment in combination with an energy budget approach recently confirmed the basic soundness of the momentum transfer theory underlying More work is needed to resolve the matter.

Physically-based models may help to elucidate the relative importance of different factors in the interception process, which are difficult to assess from an actual interception record. Two well-known models are the dynamic model developed by Rutter *et al.* (1971, 1975) and its simpler derivative, the analytical model of Gash (1979).

The Rutter model calculates a running water balance for the canopy (and trunks). It requires hourly rainfall and meteorological data as inputs, as well as values of four parameters describing the forest (free throughfall and stem-flow coefficients, canopy and trunk saturation values). In addition, it uses an empirical expression to compute drainage from the canopy. The Rutter model has been used successfully for such contrasting environments as coniferous plantations in the UK and tropical rain forest in central Amazonia.

However, it greatly underestimated E_i under wet maritime tropical conditions where unrealistically high canopy storage capacities or very low aerodynamic resistances were required to match predicted and measured E_i.

Valente *et al.* (1997) used a reformulated version of the model to predict rainfall interception by sparse canopy forests in Portugal.

The effect of varying rainfall intensities and evaporation rates, canopy storage capacity and the distribution of rain (continuous vs. intermittent) on the magnitude of E_i using the. The limiting effect of low rainfall intensity at faster evaporation rates is clearly borne out by the simulations, as is the increase in E_i at higher rainfall intensities and for higher canopy storages, especially in the case of intermittent rain.

A major practical drawback of the Rutter model is its high data requirement. The analytical model of rainfall interception developed by Gash (1979) is conceptually similar to the Rutter model but much less data-demanding.

It is usually applied on an event basis and uses average instead of hourly rates of rainfall on to, and evaporation from, a saturated canopy. Apart from evaporation from a saturated canopy, the model computes the evaporation before the canopy is fully wetted and after rainfall has ceased, the effect of small storms insufficient to saturate the canopy, and evaporation from trunks.

Total E_i is then estimated as the sum of the respective interception components. The analytical model and a derivative for sparse canopies developed by Gash *et al.* (1995) have been applied to a wide range of environments, ranging from temperate coniferous and deciduous forests to sparse Mediterranean pine and eucalypt forests and tropical rain forests. As in the case of the Rutter model, predicted amounts of E_i under more maritime conditions were underestimated by the model when evaporation from a saturated canopy was computed. For example, Gash *et al.* (1980) had to roughly double the average value of XE_{wet} to match measured and predicted E_i in a spruce plantation in Wales whereas Schellekens *et al.* (1999) required a fourfold to fivefold increase for the Puerto Rican case. A much better match of cumulative *T*/was obtained in the latter case by optimizing for XE_{wet} than for canopy storage capacity *S*.

In the Rutter and Gash models the canopy capacity is treated as a fixed parameter but in a more recent development in interception modelling (the so-called stochastic model proposed by Calder 1986) the size of the canopy capacity is allowed to vary. In the stochastic model the (repeated) wetting of the canopy is described in terms of the probability with which raindrops of different sizes strike the canopy. Larger drop sizes are considered less efficient at wetting the canopy (*i.e.* filling the storage capacity) than smaller drops.

Therefore, for the same depth of rain, larger drops will not only take longer to (re)fill the canopy capacity but also the amount of water retained on the canopy will be smaller when drops are larger. In a recent modification of the model, the wetting functions were adjusted to include the possibility of wetting of lower canopy layers by secondary drops falling from the upper layers. Although the two-layer version of the model was applied successfully in the simulation of E_i from a forest in Sri Lanka , the precise physical meaning of some of the model parameters remains somewhat vague.

In this respect it is pertinent that Uijlenhoet and Stricker (1999) recently pointed out that the stochastic model employed an erroneous rainfall intensity vs. drop size relationship. In theory, however, the stochastic approach holds considerable promise for use in humid tropical situations where raindrop sizes tend to be larger than in the temperate conditions under which the Rutter and Gash models originated. It would be interesting therefore to examine the performance of the (improved) stochastic model under the contrasting climatic conditions prevailing in central Amazonia and eastern Puerto Rico.

Models describing rates of $£_s$ (evaporation from the forest floor) range from site-specific exponential drying curves to a more explicitly physical approach that is conceptually similar to the Rutter canopy interception model.

EFFECT OF FOREST CLEARFELLING ON WATER YIELD

A basic summary of short-term results obtained from paired catchment experiments conducted before 1980 was given by Bosch and Hewlett (1982). Hornbeck *et al.* (1993) and Swank *et al.* (1988) discussed longer-term results, including the effects of multiple treatments, for catchments in the north-eastern and south-eastern USA respectively. Generally, increases in water yield during the first year after treatment are roughly proportional to percentage reductions in stand basal area, provided the latter exceed a threshold of 20-25%. Bosch and Hewlett (1982) made a distinction between coniferous and deciduous hardwood forests, suggesting that increases in flow associated with the cutting of conifers were higher (40mm per 10% change in cover) than for hardwoods (25 mm per 10% change in cover).

However, the scatter around the two tentative regression lines was large, coefficients of determination modest (at 42% and 26% respectively) and no confidence limits were provided. In a later analysis of the same dataset plus an additional 50 studies (mostly from the tropics and Australia), Sahin and Hall (1996) used fuzzy linear regression analysis and obtained the following average changes in flow per 10% change in cover: conifers, 23mm; mixed hardwoods-conifers, 22mm; and hardwoods, 17-19 mm (depending whether annual precipitation exceeded 1500mm or not). Based on this fuzzy regression approach, the overall differences between the respective vegetation groups were therefore smaller than suggested earlier by Bosch and Hewlett (1982). Stednick (1996) approached the problem of the heterogeneity in results for 95 paired catchment studies from the USA by deriving separate regression equations for each of eight geographically homogeneous regions.

In doing so, a clearer picture was obtained of the regional variation in yield increases, which ranged from about 10mm per 10% change in cover in the Rocky Mountains to over 60mm in the Central Plains. Also, some of the regionally applicable regression equations had substantially higher coefficients of determination (*e.g.* 0.65 for the Pacific Northwest or Appalachia, which both had the largest number of experiments), although correlations for other regions were extremely low (*e.g.* 0.01 for the Rocky Mountains). As such, there is something to say for the approach of Trimble *et al.* (1987) who simply pooled the coniferous and mixed hardwood data of Bosch and Hewlett (1982) and added 10 data points of their own, representing reductions in annual flow from large non-experimental ('real world') catchment areas in the Piedmont of the south-eastern USA that had undergone 10-28% reforestation.

The overall coefficient of determination for the combined dataset attained the reasonable value of 0.50 by the inclusion of the Piedmont subset, with 8 of 10 data points from the latter falling within half the standard error of the estimate (compared with only 22 of 55 data points in the Bosch and Hewlett set). This finding is all the more remarkable because half the additional data involved a change in cover of less than 20%, a value below which Bosch and Hewlett (1982) had considered effects on streamflow to be more or less non-detectable (at least on small headwater streams).

Apart from contrasts in tree species and age of the vegetation, much of the variability in the change in stream-flow observed for different sites and years after a change in cover relates to differences in climate, catchment exposure and soil depth. The climatic aspect is aptly summarized by the classic statement of Hewlett (1966) that 'it takes water to fetch water'. The importance of catchment exposure is illustrated by the difference in first-year streamflow gain after cutting differently exposed deciduous hardwood forest catchments at Coweeta in the south-eastern USA.

Flows from northerly exposed catchments increased by about 130mmyear^{-1} but increases from southerly exposed catchments could be as high as 410mm year^{-1}. As for the effect of soil depth, Trimble *et al.* (1963), when discussing differences in duration of the changes in streamflow following forest clearing in West Virginia, stated that 'the deeper the soil, the longer it takes roots of new growth or the expanding root systems of residual plants to occupy the soil mantle. When the soil mantle is reoccupied, transpiration reaches maximum levels again, and increases in streamflow disappear'. The role of soil depth can be illustrated further by examining the timing of the maximum and minimum increases in flow after forest clearance during the year. At Coweeta, where precipitation is distributed evenly over the year and the loamy soils are deep (up to 6 m) and capable of storing large amounts of moisture, the soil does not become fully recharged until early spring (April-May).

As a result, the presence or absence of a forest has little effect on the amounts of flow during this time of year (the soils being full of water anyway). However, at the onset of the growing season, increased flow due to reduced water use after clearcutting becomes evident and the contrast in streamflow from forested and cleared areas increases as the growing season advances.

The effect continues well into the winter months (November-January), even though the trees have lost their leaves by then and evaporation is strongly reduced. This reflects the development of a much larger soil water deficit below the forest during the summer that takes an additional 2 months (February-March) to be refilled by the winter rains. Conversely, on catchments with shallower or sandier soils (having less water storage and a lower water retention capacity), soil water recharge and depletion occurs much more rapidly and increases in flow after forest clearance follow the pattern of forest water use much more closely.

The effect of the timing of the cutting and the importance of controlling regrowth are demonstrated by a comparison of the results obtained for catchments 2 and 5 at Hubbard Brook (HB 2 and HB 5). On catchment HB 2, the forest was clearcut during the dormant season and regrowth eliminated by herbicide application the following spring. This produced a first-year increase in water yield of 347 mm. In contrast, on catchment HB 5, the removal of the trees took nearly a full year, while regrowth was not suppressed.

The resulting first-year increase in flow thus amounted to only 152 mm, or 44% of that observed for catchment HB 2. In addition, the gain in streamflow had largely disappeared after 3 years. Such contrasting findings once more illustrate the effect of regenerating vegetation on soil water levels and streamflow. However, it should be noted that regeneration was particularly vigorous in the example from Hubbard Brook as it took place mainly via sprouting. As demonstrated by comparative work in West Virginia, elevated streamflow levels lasted much longer (15-20 years) when regeneration had to originate from seeds because of previous herbicide applications than when sprouting was allowed (6-7 years).

Similarly, when a regenerating hardwood forest at Coweeta was cut a second time after 23 years (when streamflow was still about 80mmyear^{-1} above that observed for mature mixed forest), the initial increase in water yield was nearly identical to that of the first cut (375 vs. 362mm). However, the subsequent decline in (extra) water yield was distinctly faster. The difference was attributable to a more rapid recovery of stand biomass during the second regeneration period in association with the greater sprouting potential of the even-aged forest, which contained numerous small stems prior to the second cut. As a result, the (projected) total duration of the second period of increased flows was estimated to be 18 years shorter than that associated with the first cutting (estimated at about 35 years). Such long durations reflect the time required by the roots to reoccupy the very deep soils at Coweeta compared with those of the north-eastern USA where effects are more likely to last 3-9 years.

A rather different pattern of changes in forest water use and streamflow with time after clearing has been documented for native mountain ash *(Eucalyptus regnans)* forest in south-eastern Australia. Following clearing and burning, there is an initial rise in streamflow lasting 4-6 years. This is followed by a rapid decrease in flows until the regrowth is about 27 years old, after which run-off totals slowly return to predisturbance levels. The latter may take as long as 150 years. Vertessy *et al.* (1998) provided a mechanistic explanation for this long-term pattern in terms of concurrent changes in overstorey and understorey leaf biomass and transpiration rates, as well as changes in rainfall interception and evaporation from the forest floor.

Results from paired catchment experiments in the humid tropics also show a proportionality in flow increase with percentage of forest removed, as observed under humid temperate conditions. Reported first-year increases after clearfelling range from 125 mm in Nigeria to 820 mm in Peninsular Malaysia and show little relation to amounts of precipitation.

This led Bruijnzeel (1996) to suggest that perhaps the results were determined more by the degree of surface disturbance than by climatic or soil factors. However, in view of the large variability in total evapotranspiration among tropical forests, it may be rewarding to reanalyse the data in the light of contrasts in rainfall interception between regions as well.

Effects of Converting Natural Forest to other Land Cover Types on Water Yield

Whilst streamflow totals are observed to eventually return to preclearing levels where regrowth is allowed, the conversion of native forest to other types of vegetation cover may produce permanent changes. For example, permanent increases in annual water yield are associated with the conversion of deciduous or evergreen native forest to agricultural cropping. Reported increases range from 60-125 mm year^{-1} under humid warm temperate conditions to 450mmyear–^{1} in the equatorial tropics. The diminished water use of annual crops compared with that of a full-grown forest reflects the diminished capacity of low vegetation not only to intercept rainfall but also to extract water from deeper soil layers during periods of drought. The former relates primarily to the lesser aerodynamic roughness of short annual crops (and possibly to their smaller leaf area and thus storage capacity as well), whereas the reduced water uptake of crops reflects their more limited rooting depth. For the same reasons, conversion to pasture generally produces permanent increases in streamflow as well, although the magnitude of the increase depends on precipitation patterns and elevation. Interestingly, Hibbert (1969) noted that replacing deciduous hardwood forest in the south-eastern USA by vigorously growing *Festuca* grass did not lead to increased water yields, although increases of up to 125 mm year^{-1} were observed after the productivity of the grass declined.

Such results can be explained in part by the fact that serious water shortages that could have reduced water uptake by shallow-rooted plants do not develop in the rainy climate at Coweeta, whereas in addition the contrast between forest and grass was lessened further by the deciduous character of the forest. In contrast, the conversion of mixed deciduous hardwoods by eastern white pine *(Pinus strobus)* in the same area brought about a reduction of 250mmyear–[1] after 25 years. The decline in flow was ascribed both to a steady increase in intercepted rainfall as the pines grew older and to higher transpiration by the evergreen pine trees during the dormant season. Similarly, the higher total water use *(ET)* of mature *Pinus radiata* plantations in south-eastern Australia compared with that of native eucalypt forest has been explained in terms of the distinctly higher rainfall interception by the pines. However, where pine plantations replaced non-deciduous native forests elsewhere in the world, streamflows have been reported to return to preconversion levels within 8 years, such as in Kenya and New Zealand.

The results presented thus far pertain mostly to small headwater catchment areas involving a unilateral change in cover. Although these experiments provide a clear and consistent picture of increased water yield after replacing tall vegetation by a shorter one, and *vice versa,* such effects are more difficult to discern in large river basins having a variety of land-use types and temporal changes therein. In addition, there are the complications associated with strong spatial and temporal variability in rainfall, especially in tropical areas with largely convective rainfall, and the often large-scale withdrawals of water for municipal, agricultural and industrial purposes in populated areas. Qian (1983) was unable to detect any systematic changes in streamflow from catchments ranging in size from 7 to 727km^2 on the island of Hainan, southern China despite a 30% reduction in tall forest cover over three decades. Dyhr-Nielsen (1986) arrived at the same conclusion for the 14 500 km^2 Pasak River basin in northern Thailand, which lost 50% of its tall forest cover between 1955 and 1980.

On the other hand, Madduma Bandara and Kurupuarachchi (1988) observed an increase in averaged annual flow totals for the 1100 km^2 upper Mahaweli catchment in Sri Lanka over the period 1940-80, despite a weak negative trend in rainfall over the same period. Although both trends were not statistically significant at the 95% level, the associated increase in annual run-off ratios was highly significant.

The increased hydrological response was ascribed to the widespread conversion of tea plantations (not forest) to annual cropping and home gardens without appropriate soil conservation measures (Madduma Bandara and Kurupuarachchi 1988).

However, when analysing a longer time series (1940-97) for the 380 km^2 upper Nilwala catchment in Sri Lanka, which experienced a 35% reduction in forest cover during the observation period, a less consistent picture was obtained (Elkaduwa & Sakthivadivel 1999). Such contrasting findings illustrate the need for high-quality rainfall and streamflow data in historical hydrological data analyses, not only in the context of evaluating the effects of land-use transformations but also those associated with climatic change.

Effects of Forest Clearing on Streamflow Regimes

In areas with seasonal rainfall, the distribution of streamflow throughout the year is often of greater importance than the total annual amount *per se.* Reports of greatly diminished streamflows during the dry season after forest clearance abound in the literature, particularly in the tropics. At first sight, this seems to contradict the evidence presented earlier, that forest removal leads to higher water yields, even more so because the bulk of the increase in flow after *experimental* clearing is observed during baseflow or dry-season conditions. However, the controlled conditions imposed during catchment experiments differ from those encountered in many real-world situations.

The continued exposure of bare soil after forest clearance to intense rainfall , the compaction of topsoil by machinery or overgrazing , the gradual disappearance of soil faunal activity (Aina 1984) and increases in the area occupied by impervious surfaces such as roads and settlements all contribute to gradually reduced rainfall infiltration opportunities in cleared areas. As a result, catchment response to rainfall becomes more pronounced and the increases in storm runoff during the rainy season may become so large as to seriously impair the recharging of soil water and groundwater reserves.

When this critical stage is reached, diminished dry-season flow is the result, despite the fact that the removal of the forest has led to reduced evaporation and thus should have induced higher baseflows. On the other hand, if the surface characteristics after forest clearing are maintained sufficiently to allow the continued infiltration of (most of) the rainfall, then the effect of reduced *ET* associated with forest removal will show up as increased dry-season flow.

This could be achieved through the establishment of a well-planned and maintained road system plus the careful extraction of timber in the case of logging operations, and by applying appropriate soil conservation measures when clearing for agricultural purposes.

However, even with minimum soil disturbance, there will be local increases in stormflow volumes and peak flows after forest removal because the associated reduction in *ET* causes the soil to be wetter and thus more responsive to rainfall. Relative increases in stormflow response associated with minimum disturbance operations are largest for small events (up to 300%), declining to less than 10% for large events, although the effect will be more pronounced if soil disturbance is severe.

Also, whilst increases in stormflow response after (well-conducted) forestry operations tend to diminish or disappear altogether as the new vegetation cover becomes established, they also may become structural. Examples include run-off contributions from roads or residential areas where there were none before and permanently wetter soil conditions associated with reduced water use (*e.g.* grassland; Fritsch 1993). However, it is important to remember that caution is required when trying to extrapolate the undoubtedly adverse effects of indiscriminate forest clearing on local infiltration opportunities and storm run-off generation over large river basins.

Hydrological Effects of (re)forestation

In response to the widely observed degradation of formerly forested land and the rising demands for industrial wood and fuelwood, the need for large-scale reforestation programmes has been expressed repeatedly. It is of interest therefore to examine to what extent such plantations are capable of restoring the original hydrological conditions, *i.e.* ameliorate peak flows and, above all, enhance low flows. Whilst there is little doubt that annual water yields from forested areas are reduced compared with those for non-forested areas, it must be granted that almost no catchment forestation experiments have been carried out on seriously degraded land.

As such, it could be argued that the clear reductions in total and dry-season flows observed after afforesting natural grasslands and scrublands or pasture in South Africa , New Zealand, south-eastern Australia and Fiji only serve to demonstrate the difference in water use between the two vegetation types. In other words, the potentially beneficial effects on streamflow afforded by improved infiltration and soil water retention capacities during forest development could not become manifest in these experiments. The key question is therefore whether the reductions in storm run-off generating overland flow incurred by such soil physical improvements can be sufficiently large to compensate the extra water use by the new forest, and so (theoretically) boost low flows.

There is no easy answer to this question for several reasons. First, the effect of an increase in topsoil infiltrability on the frequency of occurrence of HOF depends equally on prevailing rainfall intensities.

For instance, rainfall intensities in the Middle Hills of Nepal were generally so low that the 140mmh^{-1} increase in infiltration capacity 12 years after reforesting a degraded pasture site with *Pinus roxburghii* made no difference in the generation of HOF. Similarly, whilst topsoil infiltrabili-ties had roughly doubled over 12 years since reforesting *Imperata* grassland with teak in Sri Lanka, at 30mmh^{-1} the hydrological impact of this increase is probably limited. A much longer time span may be required therefore to restore infiltration capacities to the high values normally found in undisturbed forests.

Secondly, the reductions in catchment response to rainfall observed after forestation may well reflect the drier soil conditions prevailing under actively growing forest rather than a reduction in peak-generating HOF.

However, none of the studies documenting decreases or increases in stormflows after land-use change has attempted to quantify the associated changes in relative contributions by subsurface and overland flow types. Third, there is also the effect of soil depth, which determines both the maximum amount of water that can be stored in a catchment under optimum infiltration conditions and the possibilities for water uptake by the developing root network of the new trees. And last but not least, there is the confounding effect of rainfall patterns (seasonal vs. well distributed) and general evaporative demand, which both exert a strong influence on tree water use.

Needless to say, the soil and climatic factors, plus the absence of detailed information on the prevailing stormflow mechanisms before and after the conversion, all render comparisons between different sites more complicated.

The only documented real-world case in which the kind of compensation mechanism supposed by Bruijnzeel (1989) seems to have occurred may be the White Hollow catchment in Tennessee, USA (Tennessee Valley Authority 1961).

Prior to improvement of its vegetation cover, two-thirds of this catchment consisted of mixed secondary forest in poor condition (due to fire, heavy logging and grazing), with another 26% under poor scrub. About 40% of the catchment was estimated to be subject to more or less severe erosion. Following extensive physical and vegetative restoration works, peak discharges decreased considerably within 2 years, especially in summer.

However, neither annual water yield nor low flows changed significantly over the next 22 years of forest recovery and reforestation. It was concluded that the extra water needed by the recovering and additionally planted trees was balanced by improved infiltration (Tennessee Valley Authority 1961). However, it is quite possible that the absence of major changes in total and low flows at White Hollow mainly reflects the lack of contrast between tall and short vegetation, as would have been the case when reforesting a truly deforested and degraded catchment.

Support for this contention comes from the observation of Trimble *et al.* (1987) that reductions in flow from several large river basins in the south-eastern USA, which had suffered considerable erosion before being partially reforested, were largest during dry years. In addition, the effect became more pronounced as the trees grew older.

A similar case was recently described for a once seriously degraded catchment in Slovenia , where the spontaneous return of forest produced a steady reduction in annual water yield over a period of 30 years, with the strongest reductions again occurring during the dry summer months.

Therefore, in both these real-world examples the water-use aspect overrides the infiltration aspect, despite the rather modest contrasts in water use between forest and grassland/cropland in the south-eastern USA.

Such findings offer little prospect for the possibility of significantly raising dry-season flows by forestation in the humid tropics. Observed maximum differences between annual water use of fast-growing pines or eucalypts and short vegetation (grass, crops) under well-watered (subtropical conditions attain values of 500-700 mm at the catchment scale and even higher values (> 1100 mm) on individual plots with particularly vigorous tree growth. The hydrological benefits incurred by the limited increases in topsoil infiltration capacities observed after forestation of degraded land in Nepal and Sri Lanka cited earlier are dwarfed by such high water requirements.

The conclusion that already diminished dry-season flows in degraded tropical areas will decrease even further upon reforestation seems inescapable.

In view of the extent of the low-flow problem, the testing of alternative ways of increasing water retention in tropical catchments without the excessive water use associated with exotic tree plantations should receive high priority.

MODELLING THE HYDROLOGICAL IMPACTS OF FOREST MANIPULATION AND LAND-USE CHANGE

As shown in the preceding sections, different types of forest manipulations affect the water (and material) flows through catchments differently. Traditionally, foresters have relied on time-consuming paired catchment experiments to evaluate such effects. Whilst this approach enabled the construction of simple nomograms from which reductions in annual streamflow after afforestation could be read as a function of stand age and annual rainfall , or first-year increases in flow as a function of percentage basal area reduction and catchment aspect , the results were often so variable as to render their applicability for more detailed water resources planning rather limited. Afforestation with eucalypts produced similarly variable responses (curves 1 and 2) that also differed markedly from those for pine. The black-box nature of the paired catchment technique is unable to evaluate the relative importance of the various factors underlying such differences and this severely limits the possibilities for extrapolation of results to other areas or periods.

Process-based hydrological models represent an alternative way of predicting how catchments might respond to different forms of management. Because many practical forest management questions have a spatial dimension to them, and because landscapes usually comprise a complex mosaic of different land uses, such models should preferably be of a 'distributed' nature, *i.e.* capable of taking into account spatial variations in topography, soils, vegetation and climate. During the last 10-15 years, considerable progress has been made in the modelling of forest hydrological and ecological processes over a range of spatial and temporal scales.

Arguably, one of the most comprehensive process-based distributed models is the topography-driven TOPOG model. TOPOG integrates the water, carbon, solute and sediment balances at the small catchment scale (typically < 10 km^2) and is thus particularly useful for exploring complex feedback mechanisms between system properties.

Examples of within-catchment applications of TOPOG relevant to forestry include the prediction of steady-state soil moisture distributions and saturated zones in the landscape during wet and dry periods, the spatial distribution of surface erosion, gully initiation and landslide hazards, and the evaluation at the hillslope scale of the water balance and growth performance of different tree planting configurations (block planting vs. strip planting) under subhumid conditions.

Off-site applications of TOPOG include the successful simulation of changes in tree growth and water yield for a small catchment after clear-felling during 20 years of regeneration.

To address forestry-related questions of catchment water management at larger scales as well, a simpler derivative of TOPOG (Macaque) has been developed recently. Macaque proved capable of effectively simulating the changes in water yield associated with the occurrence of a major wildfire (in 1939) and subsequent forest regeneration in a 163-km^2 catchment over an 82-year period (1911-92). It also enabled the prediction of the hydrological effect of the 1939 fires in a spatially explicit manner.

Models like TOPOG and Macaque constitute powerful tools whose predictions can help rational land-use decisions. However, whilst distributed models represent the only class of simulation models capable of capturing the complex feedbacks that occur upon disturbing hydrological systems, they are also data-demanding. Model outputs are especially sensitive to variations in LAI, soil hydraulic conductivity, the rainfall interception coefficient and canopy conductance, whereas at larger scales there is the problem of adequate spatial representation of rainfall inputs.

However, there is reason for optimism as various remote sensing technologies that are currently still in their infancy can be expected to become widely available within the next decade. This will greatly facilitate data acquisition and scaling-up of results.

7

Soil Electrical Conductivity and Variability

INTRODUCTION

Soil electrical conductivity (EC) is a property of soil that is determined by standardized measures of soil conductance ($resistance^{-1}$) by the distance and cross sectional area through which a current travels. Traditionally, soil paste EC has been used to assess soil salinity , but now commercial devices are available to rapidly and economically measure and map bulk soil EC across agricultural fields. The Veris® 3100 (Veris Technologies, Salina, Kansas) measures EC with a system of coulters that are in direct contact with the soil. The EM38 (Geonics, Limited, Mississauga, Ontario, Canada) induces a current into the soil with one coil and determines conductivity by measuring the resulting secondary current with another coil. Both sensors have been demonstrated to give similar results

The movement of electrons through bulk soil is complex. Electrons may travel through soil water in macropores, along the surfaces of soil minerals (*i.e.* exchangeable ions), and through alternating layers of particles and solution.

Therefore, multiple factors contribute to soil EC variability, including factors that affect the amount and connectivity of soil water (*e.g.* bulk density, structure, water potential, precipitation, timing of measurement), soil aggregation (*e.g.* cementing agents such as clay and organic matter, soil structure), electrolytes in soil water (*e.g.* salinity, exchangeable ions, soil water content, soil temperature), and the conductivity of the mineral phase (*e.g.* types and quantity of minerals, degree of isomorphic substitution, exchangeable ions). Despite the multiple causes of EC variability, bulk soil EC measurements have been related to individual factors that limit soil use and productivity such as salinity, clay content at a depth of 15-m in New Wales, Australia (r^2 = 0.78; Williams and Hoey, 1987), depth of sand deposition along the Missouri River (r^2 = 0.73-0.94; Kitchen et al, 1996), depth to claypan in Missouri

($r^2 = 0.73$; Doolittle *et al.*, 1994), and soil moisture content ($r^2 = 0.96$; Kachanoski *et al.*, 1988).

If soil EC maps have utility in production agriculture, 1) EC must be spatially structured, 2) spatial patterns must have temporal stability, and 3) EC must be related to factors of agronomic importance. The objective of this research was to determine the nature and causes of EC variability for several fields in Kentucky. Geostatistical analyses were conducted to examine the spatial and temporal variability of EC variability. Transect studies were conducted to determine the causes of EC variability.

METHODS

Site Description

This research was conducted in a field in Hardin Co, KY (Field 1), and three fields in Shelby Co, KY (Field 2, Field 3, and Field 4). Field 1 consists of the Vertrees series (Fine, mixed, mesic Typic Paleudalfs), which formed in residuum from limestone, the Nolin series (Fine-silty, mixed, mesic Dystric Fluventic Eutrochrepts), which formed in mixed alluvium, and the Crider series (Fine-silty, mixed, active, mesic Typic Hapludalfs), which formed in loess over limestone residuum. Field 2 consists of the Nicholson series (Fine-silty, mixed, mesic Typic Fragiudalfs) and the Lowell series (Fine, mixed mesic Typic Hapludalfs), both of which formed in residuum from limestone and shale and have a loess cap. Field 3 has both the Nicholson series and the Shelbyville series (Fine-silty, mixed mesic Mollic Hapludalfs), which formed in loess over residuum from limestone. Field 4 contains the Cynthiana series (clayey, mixed, mesic Lithic Hapludalfs) and Faywood series (fine, mixed, active, mesic Typic Hapludalfs), both which are shallow to bedrock and formed in residuum from limestone.

Soil EC Data Collection

A Veris® 3100 Soil EC Mapping System was used to measure soil EC. The sensor consists of six coulters, two of which introduce an electrical potential into the soil. The remaining four coulters are spaced to measure EC over two approximate depths, 0-30.5-cm ($EC_{30.5}$) and 0-91.5-cm ($EC_{91.5}$). When used in conjunction with a DGPS receiver, EC data can be geo-referenced to create a map.

Transect EC Measurements

Transects were selected from Field 1, Field 3, and Field 4. At selected points on each transect, several measurements were taken in addition to EC. Volumetric water content to a depth of 12-cm was determined with the HydroSense™ (Decagon Devices, Inc, Pullman, Washington), which uses transmission line oscillation. Depth to a clay increase was assessed using the "texture by feel" method. Penetrometer resistance was measured on all Field 1 transects. Depth measurements to fragipan (Field 3) and to bedrock (Field 4) where measured only in the fields where these factors affected crop production. Soil EC values for each point on the transect were determined by driving directly over the sampling point, stopping, and recording the $EC_{30.5}$ and $EC_{91.5}$ values. In addition, the transect EC values in Field 3 were measured on two consecutive days.

Whole Field EC Measurements

Whole field EC data were collected for a location in Hardin Co. (Field 1) on three dates: Oct. 8, 1999, March 7, 2000, and May 5, 2000. A second location in Shelby Co. (Field 2) was measured once on July 7, 1999. The fields were traversed with approximately 7.5-m between passes on each date. Field 3 was traversed in the north-south direction and the east-west direction. Data were recorded every second, and groundspeed was maintained at approximately 10.5 km hr^{-1}. Soil samples were collected for the top 15-cm using a 30.5-m grid pattern. Soil analyses, including pH, buffer pH, organic matter content, and Mehlich III extractable P, K, Ca, Mg, and Zn, were performed by the Department of Regulatory Services, at the University of Kentucky. At each sample site, all EC values falling within a 4.6-m radius were averaged and related to $EC_{30.5}$ and $EC_{91.5}$ using simple linear regression.

RESULTS AND DISCUSSION

Nature of the Variability

Day-to-day variability was greater for $EC_{30.5}$ than for $EC_{91.5}$. While collecting EC data, extreme values were encountered on occasion, as can be seen by the high value for May 4th $EC_{91.5}$. The small-scale temporal variability also gives an indication of measurement error. The relative nugget variances for both depths also give an indication of measurement error ($EC_{30.5}$, 46%; $EC_{91.5}$, 27%) and were large.

The larger scale temporal variability reflects changes in EC associated with different environmental conditions. While the EC values were substantially lower during the drought of 1999 (October 8th, 1999) than in the spring of 2000, the general spatial patterns in EC were similar across all three dates.

Both $EC_{30.5}$ and $EC_{91.5}$ tended to be higher on the Vertrees series, which is located in the lower right, lower left, and upper right regions of Field 1. The values tended to be lower for the Nolin series, which are mainly located in the depressions. The remainder of the field was the Crider series. Soil EC was a good indicator of soil type for this field. The Vertrees soils have a red, Bt horizon near the surface, which increases conductivity. The EC values for the Vertrees series were not as high on October 8th, 1999 (during the drought) as on the later two dates. This suggests that soil moisture enhances the conductivity of clay. This did not, however, greatly change the overall appearance of the maps on the different dates because each was based on an equal number of observations in each category rather than equal sizes of mapping intervals. Date of the measurement did, however, change the spatial statistics of the data. Anisotropic behavior was not apparent on October 8th, 1999, but it was on the latter two dates.

The dark diagonal lines indicate the direction of the anisotropic axis with of minimum spatial variability (here the northwest-southeast direction). Orthogonal to this direction indicates the direction with the maximum spatial variability. Anisotropy was present in this field because the Vertrees soil, which occurred in southwest, northeast, and southeast regions of the field, had very large EC values on the latter two dates when soil moisture was greater. Therefore, variability was much greater in the southwest-northeast direction. In all cases, the anisotropy does not seem to be a great issue within the first 50-m. Anisotropy was less important with $EC_{91.5}$. This may be because at greater depths, the soil was more uniform in clay and moisture content. The impact of anisotropy in this field would depend upon the depth of interest, whether the data would be used for interpolation, and the sampling interval.

FACTORS AFFECTING SOIL ANALYSIS

Nutrient values can vary as a result of a number of factors which are discussed below. Further study is needed to define the scale and cause of such variation.

SPATIAL VARIATION

Variation can exist over very short distances (less than 1 metre) and the number of cores that are needed to ensure that sampling is representative of an area will vary according to the scale of this variation. When sampling to obtain an average value for a field this is not so important because all the cores taken are bulked into a single sample; for a typically uniform field 25 cores are sufficient.

However, local variation is important when grid sampling because each grid point is viewed as a separate sample. Normally, because the extent of the variation is unknown, at least 16 cores need to be taken on 1 metre grid-spacing around each grid point to provide a single bulked sample for analysis. If it is planned to produce a 'map' of the nutrient variability within the field, then the grid points should be not more than 50 metres apart. Where considerable differences occur within a field, a single sample can be misleading because the averaging will disguise different treatment requirements. If different areas are known to vary they should be sampled separately. Knowledge of field amalgamations, soil type changes observed when ploughing or cultivating, visual crop growth difference across the field or any other specific information on field variation should be used to ensure that the area is sampled to best effect.

Temporal Variation

Variations in nutrient values have been observed at different times of the year and there is evidence to suggest that soil P, K and Mg values will be higher in the early spring than in the autumn as a result of chemical weathering over winter, biological activity and lack of uptake by growing crops

Moisture

If soils are dry at the time of sampling the analytical results can be affected and may appear a little lower for pH, P and possibly K. Movement and uptake of all nutrients will of course be restricted in very dry soils but this is a transient problem and does not reflect the normal availability of nutrients in the soil. At present it is not possible to quantify this effect in order to improve interpretation of analytical results.

Crop Removal

During periods of rapid growth crop uptake, especially of potassium, can be large and may deplete available soil nutrient levels for a short period until the nutrient status returns to equilibrium. This could affect results for some soils if samples are taken at such times.

Nutrients are returned to the soil in crop residues but will not be determined by analysis until the plant material is broken down. It has been suggested that sampling close to harvest may result in an under-estimation of true soil nutrient status.

Sampling Depth

Frequently there may be a gradient in nutrient level down the soil profile, usually declining with depth, reflecting the accumulation of nutrients in the plough layer. This is accentuated in minimal cultivation systems where phosphate and potash may be concentrated in the top 2 or 3 inches (5 to 8 cm). With continuous direct drilling there may be a large difference between the top 2 inches (5 cm) and 3 to 6 inch (8 to 15 cm) zones. For permanent grass where there is no soil disturbance, consistency of sampling depth of 3 inches (8 cm) is particularly important.

Previous Manuring

Applications of fertilisers and manures obviously have a major impact on measurement of soil nutrients. It is difficult to lay down rules as to how soon sampling should be undertaken after application. General guidelines are as follows:

8 weeks after P, K, Mg fertiliser application.12 weeks after slurry or manure application.12 months after lime spreading.

Ploughing and cultivating help distribute nutrients from fertilisers and manures, and lime, throughout the depth of cultivation, but this takes time.

SAMPLING

Soil analysis data are only as good as the sample taken. A sample normally comprises around 1 kg (2 lbs) of soil which is taken to represent an entire area or field, which contains around 2,000 t of soil per hectare to a plough depth of 8 inches (20 cm).

It is therefore imperative to obtain as representative a sample as possible or the results will not reflect the nutrient status accurately.

Rules of Sampling

1. Use of a suitable tool (cone auger, screw auger, corer etc) which facilitates and encourages the taking of more rather than fewer cores, of a uniform size and down to the full depth of sampling.
2. Use equipment and packaging that will not contaminate the sample. Galvanised sampling tools are unacceptable for trace element analysis.
3. Label samples clearly.
4. Sample to a consistent depth. Normal depth is 6 inches (15 cm) for arable soils, 3 inches (7.5 cm) for grassland.
5. Divide the field into areas which are as uniform as possible in soil type, past cropping, and manuring history and sample separately. Small areas of different soil *e.g.* wet, chalky, shallow, stony etc. should be excluded.
6. Avoid headlands, gateways, trees, mole hills, dung/urine patches, water troughs, areas where lime or manure has been dumped, old hedgerows/middens/ponds or any other irregular feature.
7. Discard stone and plant debris.
8. Take at least 25 cores from each area to be sampled and put them together to form a single representative sample. The numbers of cores should not be restricted simply because the container is full! Thoroughly mix all cores and take a sub-sample from this for despatch to the laboratory-this must be done carefully.
9. Ensure the sample represents the whole area. Sample on a W pattern over the field; for a regular shaped field this means 7 cores per leg of the "W".
10. In the case of grid sampling up to 16 cores are needed at each sampling point to obtain a representative sample. These should be taken in a regular pattern about 1 metre apart around the grid point. Grid points should be evenly spaced over the field and should not be more than 50 metres apart if a map is to be produced.

11 Sample at the same point in the rotation, before the crop which is most demanding or responsive to P and K. In descending order of importance these are: horticultural crops, vegetables, roots, pulses, spring sown combinable crops, winter sown combinable crops. For pH it is preferable to sample 12 months before a sensitive crop such as sugar beet or barley.
12 Sample at the same time of the year.
13 Avoid sampling under extremes of soil conditions *e.g.* waterlogged or very dry soil.
14 Do not sample within 8 weeks of fertilising, or within 12 weeks of manure or slurry application, for P, K and Mg analysis or sooner than 12 months after liming for pH analysis.
15 Maintain records and use the analytical results to develop nutrient management plans.
16 Where sampling to diagnose a crop problem take multi-cored samples (at least 16 cores per bulked sample) from areas of poor growth and separate samples from normal (good) areas. The relative values between good and poor will be more informative than the actual values of the problem area.

Frequency of Sampling

Soil nutrient levels do not alter markedly over short periods of time unless major factors of supply or demand are introduced. There is therefore no point in incurring the additional cost of sampling more frequently than necessary, except perhaps as a check to ensure that earlier sampling has been carried out properly. Additional sampling may be justified when there is a major change to husbandry practice-for instance alteration of cropping or manure policy. General guidelines for sampling frequency are as follows:-

	Sample every
Permanent grass	7 years
Intensively used grassland	3-4 years
General arable cropping	4-5 years
Arable/grass systems	4-5 years
Field vegetables & horticulture	2-3 years

Analysis

There are a number of reasons why different results may be obtained for identical samples analysed by different laboratories. The most obvious reason is that different extractants may have been used. Variations in results can also occur if identical protocols are not followed-techniques of soil drying, grinding and sieving, reagent and equipment temperature, extractant concentrations, extraction shaking, stirring and filtration, and extraction time can all affect the result.

Results are usually reported in mg/l but where a laboratory measures soil by weight instead of volume, the units will be as mg/kg. This can give rise to differences on some soils-particularly those with high levels of organic matter.

If different laboratories are used for analyses these factors must be taken into account.

INTERPRETATION OF SOIL PH

In England & Wales target pH (measured in water) is 6.5 for arable crops and 6.0 for grassland. Optimum range in Scotland where Scottish Agricultural College (SAC) measure pH in calcium chloride is 6.0-6.5 for arable crops and 5.7-6.2 for grassland.

In-field measurements using pH indicator on some soils where free chalk or lime particles exist may give lower values than laboratory results for the same field. This is because grinding the soil for laboratory analysis pulverises any chalk/lime particles and the pH as measured is increased. Acidity below pH 6.0 will reduce the availability of nutrients, especially P.

Soil analysis may overstate the availability of K on some alkaline soils where pH is over 7.0 and responses to added potash may occur at higher soil K levels than would be expected.

Availability of trace elements is radically affected by pH and the need for trace elements should only be assessed after any required amendment of acidity has been undertaken and has had time to take effect.

8

Climate Change Impacts on Forest Strength

INTRODUCTION

The world's climate is changing. Increased temperatures and levels of atmospheric carbon dioxide as well as changes in precipitation and in the frequency and severity of extreme climatic events are just some of the changes occurring. These changes are having notable impacts on the world's forests and the forest sector through longer growing seasons, expansion of insect species ranges, and increased frequency of forest fires. Climate change can affect forest pests and the damage they cause by: directly impacting their development, survival, reproduction and spread; altering host defences and susceptibility; and indirectly impacting the relationships between pests, their environment and other species such as natural enemies, competitors and mutualists. A deeper understanding of the complex relationships between a changing climate, forests and forest pests is vital to enable those in forest health protection and management to expect and prepare for changes in pest behaviours, outbreaks and invasions. This chapter investigates these relationships and the implications for forest health protection and management.

Specific examples where insect and pathogen lifecycles or habits have been altered by local, national or regional climatic changes are discussed. For example, the mountain pine beetle (*Dendroctonus ponderosae*) has shown decreases in generation time and winter mortality resulting in exponential population growth and major range extension in western North America. Warmer temperatures have resulted in range expansions of pests such as the pine processionary caterpillar (*Thaumetopoea pityocampa*) and the oak processionary caterpillar (*T. processionea*) in Europe and the southern pine beetle (*D. frontalis*) and red band needle blight (*Mycosphaerella pini*) in the US. Similar patterns are expected for other pathogens such as *Armillaria mellea*, and

Phytophthora cinnamomi in Europe. In the Democratic Peoples Republic of Korea, decreased winter temperature and lighter snow cover have contributed to reduced mortality and early emergence of overwintering *Dendrolimus spectabilis* (pine moth) with resulting loss of native *Pinus densiflora.*

The purpose of this report is to compile and summarize the present knowledge on the impacts of climate change on forests and forest pests. Since limited research has been dedicated to determining such impacts on forest pests in particular, information on non-forest pests is provided throughout the report to enable a better understanding of the potential impacts of climate change on forest health.

CURRENT KNOWLEDGE AND FUTURE EXPECTATIONS

The world's climate is changing. While there are natural climate variations, the climate change we are most concerned about is in regards to the modifications to the greenhouse effect that human activities have caused. The Intergovernmental Panel on Climate Change (IPCC), in its Fourth Assessment Report, concluded with more certainty that global climate change is unequivocal and it is widely believed to result primarily from the effects of emissions of carbon dioxide (CO_2) and other greenhouse gases (GHGs) such as methane (CH_4) and nitrous oxide (N_2O), from human activities.

This conclusion was based on a number of observations about the Earth's climate including the following.

- Global surface temperature has increased by an estimated 0.74 degrees Celsius (°C) over the past century. Over a 50 year period from 1956 to 2005 the warming trend was nearly twice that for the 100 years from 1906 to 2005. Eleven of the 12 years from 1995 to 2006 rank among the 12 hottest years on record (since 1850, when sufficient worldwide temperature measurements began). Increases are widespread globally and are greater at higher northern latitudes. Over the last 50 years, cold days and nights as well as frosts have become less frequent over most land areas, while hot days and nights and heat waves have become more frequent.
- Consistent with warming, the extent of snow and ice has decreased. Mountain glaciers and snow cover have also declined on average worldwide. The maximum area of seasonally frozen ground in the Northern Hemisphere has decreased by about 7 percent since 1900, with decreases in the spring of up to 15 percent. Satellite data since 1978 show that the extent of Arctic sea ice during the summer has shrunk by more than 20 percent.
- Since 1961, the world's oceans have been absorbing more than 80 percent of the heat added to the climate, causing ocean water to expand thereby contributing to rising sea levels. This expansion was

the largest contributor to sea level rise between 1993 and 2003. Melting glaciers and losses from the Greenland and Antarctic ice sheets have also contributed to recent sea level rise. The incidence of extreme high sea level has increased at a number of sites globally since 1975.

- From 1900 to 2005, significant increases in precipitation have been observed in eastern parts of North and South America, northern Europe and northern and central Asia. The frequency of heavy precipitation events has also increased over most areas. In contrast, precipitation has declined in the Sahel, the Mediterranean, southern Africa and parts of southern Asia. Droughts have become longer and more intense worldwide and have affected larger areas since the 1970s, particularly in the tropics and subtropics.
- Evidence of an increase in intense tropical cyclone activity in the North Atlantic has been observed since about 1970 and there are suggestions of similar increased activity in some other regions.

PREDICTIONS FOR THE FUTURE

The IPCC has made a number of predictions for future climate change. A warming of about 0.2°C per decade is projected for the next two decades; temperature projections beyond this period depend on specific emissions scenarios. Even if the concentrations of all GHGs and aerosols were kept constant at year 2000 levels, a further warming of about 0.1°C per decade would be expected. The full range of projected temperature increase, based on six emission scenarios, is 1.1-6.4 °C by the end of the century. The best estimate range of projected temperature increase, which extends from the midpoint of the lowest emission scenario to the midpoint of the highest, is 1.8-4.0 °C by the end of the century.

It is very likely that continued GHG emissions at or above current rates would cause further warming and induce many changes in the global climate system during the 21st century that would be larger than those observed during the 20th century. Geographically the predicted warming trends for the 21st century are believed to be similar to those observed over the past several decades whereby temperature increases are expected to be greatest over land and at most high northern latitudes, and least over the Southern Ocean (near Antarctica) and the northern North Atlantic. It is very likely that extreme heat, heat waves and heavy precipitation events will become more frequent. Increases in high latitude precipitation are very likely, while decreases are likely in most subtropical regions such as Egypt. It is also likely that future tropical cyclones (typhoons and hurricanes) will become more intense, with higher peak wind speeds and heavier precipitation associated with warmer tropical seas. Snow cover area is projected to contract, widespread increases in thaw depth are projected over most permafrost regions and sea ice is projected to shrink in both the Arctic and Antarctic. In some projections, Arctic late-summer sea ice disappears almost entirely by the latter part of the 21st century.

Climate change is impacting the world's ecosystems and it is expected that the magnitude of these impacts will increase along with temperatures over this century. Many species and ecosystems may not be able to adapt as the effects of global warming and associated disturbances, such as floods, drought, wildfire and insect outbreaks, are compounded by other stresses such as land use change, overexploitation of resources, pollution and fragmentation of natural systems. If the global average temperature increases more than 1.5-2.5 °C, it is believed likely that approximately 20-30 percent of plant and animal species assessed so far will be at an increased risk of extinction. Major changes in ecosystem structure and function, species' ecological interactions and species' geographical ranges, with predominantly negative consequences for biodiversity and ecosystem goods and services are also projected.

IMPACTS ON FORESTS AND THE FOREST SECTOR

Climate change, in particular increased temperatures and levels of atmospheric carbon dioxide as well as changes in precipitation and in the frequency and severity of extreme climatic events, is having notable impacts on the world's forests and the forest sector.

PRODUCTIVITY AND HEALTH

Forest productivity and species diversity typically increase with increasing temperature, precipitation and nutrient availability, although species may differ in terms of their tolerance. As a key factor that regulates many terrestrial biogeochemical processes, such as soil respiration, litter decomposition, nitrogen mineralization and nitrification, denitrification, methane emission, fine root dynamics, plant productivity and nutrient uptake, temperature changes are likely to drastically alter forests and ecosystem dynamics in many ways. The impacts of elevated temperatures on trees and plants will vary throughout the year since warming may relieve plant stress during colder periods but increase it during hotter periods.

Moisture availability in forests will be strongly influenced by changes in both temperature and precipitation. Warmer temperatures lead to increased water losses from evaporation and evapotranspiration and can also result in reduced water use efficiency of plants. Longer, warmer growing seasons can intensify these effects resulting in severe moisture stress and drought. Such conditions can lead to reductions in the growth and health of trees although the severity of the impacts depends on the forest characteristics, age-class structure and soil depth and type. Young small plants such as seedlings and saplings are particularly susceptible whereas large trees with a more developed rooting system and greater stores of nutrients and carbohydrates tend to be less sensitive to drought, though they are affected by more severe conditions. Shallow-rooted trees and plants as well as species growing in shallow soils are more susceptible to water deficits. Deep-rooted trees can absorb water from greater depths and therefore are not as prone to water stress. Moisture stress and drought can also impact forest health by enhancing susceptibility to disturbances such as insect pests and pathogens and forest fires.

Higher atmospheric CO_2 levels result in increased growth rates and water use efficiency of plants and trees, so long as other factors such as water and nutrients (*e.g.* nitrogen, phosphorus, sulphur, some micronutrients) do not become limiting. It has been suggested however that this positive effect declines with increasing concentrations. However current free air CO_2 enrichment facilities have observed multi-year growth increases of 23 percent with a 175 ppm enrichment above a 375 ppm ambient CO_2 concentration. Elevated carbon dioxide levels can also result in changed plant structure such as increased leaf area and thickness, greater numbers of leaves, higher total leaf area per plant, and larger diameter stems and branches. It is important to note that plant responses to CO_2 enrichment may differ between species and local environmental conditions which are likely to result in substantial changes in the species composition and dynamics of terrestrial ecosystems. Concurrent increases in concentrations of ground-level ozone (O_3) may lower tree productivity and enhance susceptibility to pathogens , while N_2O may enhance growth in nitrogen-limited ecosystems such as boreal forests.

RESPONSES OF SPECIES AND COMMUNITIES TO CLIMATE CHANGE

Consistent responses of species and communities to climate change or 'fingerprints' are typically associated with changes in their distribution, particularly at their latitudinal or altitudinal extremes. The distribution of forest plants and trees is expected to shift northwards or to higher altitudes in response to climate warming (Parmesan, 2006; Menéndez, 2007). In a recent study, Lenoir *et al.* (2008) compared the altitudinal distribution of 171 forest plant species between two periods 1905-1985 and 1986-2005 in Western Europe and concluded that climate warming has resulted in a significant upward shift in species optimum elevation (altitude of maximum probability of presence) averaging 29 m per decade. This study showed that climate change affects not just species' ranges at their distributional margins but at the spatial core of the distributional range of plant species. The quickest species noted to relocate were those with shorter life spans and faster reproduction cycles such as herbs, ferns and mosses; larger long-lived trees and shrubs did not show a significant shift and are thus under greater threat from the impacts of climate change because they can't adapt to local conditions quickly enough and relocate. Such distributional changes will no doubt result in forest ecosystems that are very different from what we see today. A similar study from 26 mountains in Switzerland reported that alpine flora have expanded toward the summits since the 1940s. Upward movements of tree lines have also been observed in Siberia, the Canadian Rocky Mountains, and New Zealand and northward shifts have been noted in Sweden and Eastern Canada.

The timing of such shifts however will not be solely determined by temperature but will depend on a number of factors such as the rate at which seeds can disperse into new regions that are climatically suitable (*i.e.* with proper moisture conditions, soil characteristics and nutrient availability), possible human interventions to promote movement of species, and changes to disturbance regimes (Shugart, Sedjo and Sohngen, 2003).

VARIETY OF DISTURBANCES

Forests are subjected to a variety of disturbances that are themselves strongly influenced by climate. Disturbances such as fire, drought, landslides, species invasions, insect and disease outbreaks, and storms such as hurricanes, windstorms and ice storms influence the composition, structure and function of forests. Climate change is expected to impact the susceptibility of forests to disturbances and also affect the frequency, intensity, duration, and timing of such disturbances. For example, increased fuel loads, longer fire seasons and the occurrence of more extreme fire weather conditions as a consequence of a changing climate are expected to result in increased forest fire activity. A changing climate will also alter the disturbance dynamics of native forest insect pests and pathogens, as well as facilitating the establishment and spread of non indigenous species. Such changes in disturbance dynamics, in addition to the direct impacts of climate change on trees and forest ecosystems, can have devastating impacts particularly because of the complex relationships between climate, disturbance agents and forests.

Either of these disturbances can result in forest susceptibility to other disturbances. For example, pine forests in Central America became infested with bark beetles, primarily *Dendroctonus frontalis* in association with other *Dendroctonus* and *Ips* species, after suffering damage from Hurricane Mitch in 1998. The beetle outbreaks caused further extensive tree mortality thereby increasing fuel loads in the region's forests which severely increased the risk of wildfires. Making predictions on the future impacts of a changing climate on forest disturbances is made more difficult by these interactions.

Impacts on Trees and Forests

All of these impacts on trees and forests will inevitably have widespread impacts on the forest sector. Changes in the structure and functioning of natural ecosystems and planted forests (due to temperature changes and rainfall regimes) and extreme events and disasters (hurricanes, droughts, fires and pests) will have negative impacts on the productive function of forest ecosystems which in turn will affect local economies. Production patterns and trade in forestry commodities will be altered as species are grown more competitively in higher latitudes and altitudes.

Conversely, markets may be saturated due to increased mortality of trees following pest infestations as has been experienced with the mountain pine beetle in Canada. Decreased forest ecosystem services, especially water cycle regulation, soil protection and conservation of biological diversity, as a result of climate change may imply increased social and environmental vulnerability. While climate change is likely to increase timber production and lower market prices in general, the increases in production will certainly not be evenly distributed throughout the world; some areas will experience better conditions than others (Pérez-García *et al.*, 2002). For example, forests with low productivity due to drought will likely face further decreases in productivity, while areas where temperature limits productivity may benefit from rising temperatures.

IMPACTS ON INSECTS AND DISEASES

Changes in the patterns of disturbance by forest pests (insects, pathogens and other pests) are expected under a changing climate as a result of warmer temperatures, changes in precipitation, increased drought frequency and higher carbon dioxide concentrations. These changes will play a major role in shaping the world's forests and forest sector. Climate change can affect forest pests and the damage they cause by: directly impacting their development, survival, reproduction, distribution and spread; altering host physiology and defences; and impacting the relationships between pests, their environment and other species such as natural enemies, competitors and mutualists.

Direct Impacts

Climate, temperature and precipitation in particular, have a very strong influence on the development, reproduction and survival of insect pests and pathogens and as a result it is highly likely that these organisms will be affected by any changes in climate. With their short generation times, high mobility and high reproductive rates it is also likely that they will respond more quickly to climate change than long-lived organ-isms, such as higher plants and mammals and thereby may be the first predictors of climate change.

Physiology

Climate influence on insects can be direct, as a mortality factor, or indirect, by influencing the rate of growth and development. Some information on the impacts of increased CO_2, and O_3, is becoming available but only for specific environments and only very partial information is available on changing UVB levels and altered precipitation regimes. For these reasons this report will focus on the impacts of temperature. Temperature is considered to be the more important factor of climate change influencing the physiology of insect pests. Precipitation however can be a very important factor in the epidemiology of many pathogens, such as *Mycosphaerella pini,* that depend on moisture for dispersal.

The magnitude of the impacts of temperature on forest pests will differ among species depending on their environment, life history, and ability to adapt. Flexible species that are polyphagous, occupy different habitat types across a range of latitudes and altitudes, and show high phenotypic and genotypic plasticity are less likely to be adversely affected by climate change than specialist species occupying narrow niches in extreme environments.

Increases in summer temperature will generally accelerate the rate of development in insects and increase their reproductive capacity while warmer winter temperatures may increase overwinter survival (Ayres and Lombardero, 2000; Logan, Régnière and Powell, 2003). Perhaps the best example of such impacts is the mountain pine beetle (*Dendroctonus ponderosae*) which has been at epidemic proportions in western Canada for several years. Successive years of mild winters have decreased the mortality of overwintering stages and generation time allowing for massive destruction of pines in the region, particularly lodgepole pine (*Pinus contorta*). Decreased snow depth associated with warmer winter temperatures may also decrease the winter survival of many forest insects that overwinter in the forest litter where they are protected by snow cover from potentially lethal low temperatures.

The impact of a change in temperature will vary depending on the climatic zone. In temperate regions, increasing temperatures are expected to decrease winter survival while in more northern regions, higher temperatures will extend the summer season thereby increasing growth and reproduction. Given the more severe environmental control, and greater predicted increases in temperature in boreal and polar regions, the impacts of temperature are expected to be greater on species from those regions than on species in temperate or tropical zones. However, Deutsch *et al.* (2008) suggest that, in the absence of ameliorating factors such as migration and adaptation, the greatest extinction risks from global warming may be in the tropics. Warming in the tropics, though proportionately smaller in magnitude, could have the most deleterious impacts because tropical insects have very narrow ranges of climatic suitability compared to higher latitude species, and are already living very close to their optimal temperature.

Some important forest insect pests have critical associations with symbiotic fungi but limited information is available on how temperature changes may affect these symbionts and thus indirectly affect host population dynamics. In some cases insect hosts and their symbionts may be similarly affected by climatic change while in other cases, hosts and symbionts may be affected asymmetrically, effectively decoupling the symbiosis.

Distribution Limits of a Species

Climate plays a major role in defining the distribution limits of a species. With changes in climate, these limits are shifting as species expand into higher latitudes and altitudes and disappear from areas that have become climatically unsuitable.

Such shifts are occurring in species whose distributions are limited by temperature such as many temperate and northern species. It is now clear that poleward and upward shifts of species ranges have occurred across many taxonomic groups and in a large diversity of geographical locations during the 20th century. Parmesan and Yohe (2003) reported that more than 1 700 Northern Hemisphere species have exhibited significant range shifts averaging 6.1 km per decade towards the poles (or 6.1 m per decade upward).

The range expansions of many lepidopterans have been particularly well documented. Parmesan *et al.* (1999) reported a poleward shift of 35-240 km for 22 out of 35 non-migratory European butterfly species during the last century.

Wilson *et al.* (2005) noted that the lower elevational limits of 16 butterfly species in central Spain had risen on average by approximately 212 m in 30 years, a rise attributed to an observed 1.3 °C rise in mean annual temperature. Wilson *et al.* (2007) showed uphill shifts of approximately 293 m in butterfly communities in the Sierra de Guadarrama of central Spain between 1967-1973 and 2004-2005 as a result of climate warming. Climate change may also weaken the association between climatic and habitat suitability. Franco *et al.* (2006) concluded the importance of climate warming and habitat loss in driving local extinctions of northern species of butterflies in northern Great Britain over the past few decades.

Forest pests are also occurring outside historic infestation ranges and at intensities not previously observed. Some examples of forest pest species that have responded or are predicted to respond to climate change by altering distribution include the following.

- A major epidemic of the mountain pine beetle (*Dendroctonus ponderosae*) has been spreading northwards and upwards in altitude in western Canada (British Colombia and more recently, Alberta) for several years.
- Warmer temperatures have influenced the southern pine beetle (*D. frontalis*) resulting in range expansions in the United States.
- The pine processionary caterpillar (*Thaumetopoea pityocampa*) has significantly expanded its latitudinal and altitudinal distribution in Europe.
- The oak processionary caterpillar (*T. processionea*) has shifted its distribution north in Europe during the latter half of the 20th century.
- The European rust pathogen *Melampsora allii-populina* is likely to spread northwards with increased summer temperatures.

- The root rot pathogen *Phytophthora cinnamomi* is predicted to spread into colder regions of Europe and have increased severity with climate change scenarios of increased average temperatures.

The ability of a species to respond to global warming and expand its range will depend on a number of life history characteristics, making the possible responses quite variable among species. Bale *et al.* (2002) suggested that fast-growing, non-diapausing insect species or those not dependent on low temperature to induce diapause, will respond to warming by ex-panding their distribution whereas slow-grow-ing species which need low temperatures to induce diapause (*i.e.* boreal and mountain species in the northern hemisphere) will suffer range contractions. Range-restricted species, particularly polar and montane species, show more severe range contractions than other groups and are considered most at risk of extinction due to recent climate change. Range shifts may be limited by factors such as day length or the presence of competitors, predators or parasitoids. For example, the range expansion of insects which are very host-specific (specialists) may be limited by the slower rate of spread of their host plant species (Harrington, Fleming and Woiwod, 2001).

PHENOLOGY: TIMING OF SEASONAL ACTIVITIES OF PLANTS AND ANIMALS

Phenology is the timing of seasonal activities of plants and animals such as flowering or breeding. Since it is in many cases temperature dependent, phenology can be expected to be influenced by climate change. It is one of the easiest impacts of climate change to monitor and is by far the most documented in this regard for a wide range of organisms from plants to vertebrates. Common activities to monitor include earlier breeding or first singing of birds, earlier arrival of migrant birds, earlier appearance of butterflies, earlier choruses and spawning in amphibians and earlier shooting and flowering of plants. Evidence of phenological changes in numerous plant and animal species as a consequence of climate change is abundant and growing. In general, spring activities have occurred progressively earlier since the 1960s and has been documented on all but one continent and in all major oceans for all well-studied marine, freshwater, and terrestrial groups.

Where life cycle events are temperature-dependent, they may be expected to occur earlier and increased temperatures are likely to facilitate extended periods of activity at both ends of the season, provided there are no other constraints present (Harrington, Fleming and Woiwod, 2001). With increased temperatures, it is expected that insects will pass through their larval stages faster and become adults earlier. Therefore expected responses in insects could include an advance in the timing of larval and adult emergence and an increase in the length of the flight period. Members of the Order Lepidoptera again provide the best examples of such phenological changes. Changes in butterfly phenology have been reported from the UK where 26 of 35 species have advanced their first appearance. First appearance for 17 species in Spain has advanced by 1-7 weeks in just 15 years. Seventy percent of 23 butterfly species in California, USA have seen an advancement of first flight date of approximately eight days per decade. Changes in phenology - early adult emergence and an early arrival of migratory species - have also been noted for aphids in the UK (Zhou *et al.*, 1995; Harrington *et al.*, 2007). Gordo and Sanz (2005) investigated climate impacts on four Mediterranean insect species (a butterfly, a bee, a fly and a beetle) and noted that all species exhibited changes in their first appear-ance date over the last 50 years which was correlated with increases in spring tem-perature.

Parmesan and Yohe (2003) estimated that more than half (59 percent) of 1598 species investigated exhibited measurable changes in their phenologies and/or distributions over the past 20-140 years. They also estimated a mean advancement of spring events by 2.3 days/decade based on the quantitative analyses of phenological responses for these species. Root *et al.* (2003), in a similar quantitative study, estimated an advancement of 5.1 days per decade. Parmesan (2007) investigated the discrepancy between these two estimates and noted that once the differences between the studies in selection criteria for incorporating data was accounted for, the two studies supported each other, with an overall spring advancement of 2.3-2.8 days/decade found in the resulting analysis. However, in this last study, latitude explained only 4 percent of overall variation of phenological changes even though it is strongly associated with the importance of warming trends. This last observation may relate to the importance of the change in climate relative to the natural amplitude of the climate variability.

INDIRECT IMPACTS OF CLIMATE CHANGE ON HOST TREES

Changes in temperature, precipitation, atmospheric CO_2 concentrations and other climatic factors can alter tree physiology in ways that affect their resistance to herbivores and pathogens.

Drought : The Most Important Climate

Drought is one of the most important climate-related events through which rapid ecosystem changes can occur as it affects the very survival of existing tree populations. Long-term drought can result in reduced tree growth and health thereby increasing their susceptibility to insect pests and pathogens. A number of insect pests and diseases are associated with stressed trees, such as *Agrilus* beetles and the common and widespread *Armillaria* species which have been linked to oak decline. Others are limited by host defences in healthy trees, such as the European spruce bark beetle (*Ips typographus*). Drought can also elicit changes in plant and tree physiology which will impact pest disturbance dynamics. Leaves may change colour or become thicker or waxier which could affect their palatability to insects (Harrington, Fleming and Woiwod, 2001). The concentration of a variety of secondary plant compounds tends to increase under drought stress which would also lead to changes in the attraction of plants to insect pests (Harrington, Fleming and Woiwod, 2001). Moderate drought however may actually increase production of defence compounds in plants and trees possibly providing increased protection against pests.

Sugar concentrations in foliage can increase under drought conditions making it more palatable to herbivores and therefore resulting in increased levels of damage (Harrington, Fleming and Woiwod, 2001). Increases in the sugar content in drought-stressed balsam fir for example have been known to stimulate the feeding of certain stages of spruce budworm (*Choristoneura fumiferana*) and accelerate their growth. Another advantage for forest pests is the increased temperature of drought-stressed trees, which can be 2-4 °C warmer, which can benefit the fecundity and survival of insects for example. The impacts of such changes to host tree physiology and susceptibility provoke different responses from pest species. Rouault *et al.* (2006) investigated the impacts of drought and high temperatures on forest insects and noted that woodborers were positively influenced by the high temperatures which increased their development rates and the prolonged water stress that lowered host tree resistance while defoliators benefited from the increased nitrogen in plant tissues linked to moderate or intermittent water stress.

The large natural spatial and temporal variability in forest processes makes it difficult to positively relate drought-related tree mortality to a greater incidence of pest or fungal pathogen damage. In a recent study on the impact of a large-scale, multi-annual drought on the growth and mortality of aspen (*Populus tremuloides*) in Canada, Hogg, Brandt and Michaelian (2008) could not find a significant relationship between drought severity and either insect defoliation or wood borer infestations. Drought severity was, however strongly related to total growth loss and mortality, but the delay of mortality into the years following the drought suggested that secondary agents may have been involved in the process.

ELEVATED LEVELS OF ATMOSPHERIC CARBON DIOXIDE

The atmospheric CO_2 levels result in improved growth rates and water use efficiency of plants and trees. This increased productivity leads to lower nitrogen concentrations in trees and plants as carbon-nitrogen (C: N) ratios rise and thus reduces the nutritional value of vegetation to insects (Kopper and Lindroth, 2003; Mortsch, 2006). In response insects may increase their feeding (and consequently tree damage) in an attempt to compensate for the reduced quality and gain the necessary nitrogen. In many cases the increased feeding does enable the insect to meet its nutritional needs but most often it does not and results in poor performance, reduced growth rates and increased mortality. Such an effect, however, is not consistently observed (Holton, Lindroth and Nordheim, 2003), and increased growth due to enhanced CO_2 may in fact more than compensate for the defoliation in some cases (Kopper and Lindroth, 2003).

Elevated CO_2 levels can also result in changed plant structure such as increased leaf area and thickness, greater numbers of leaves, higher total leaf area per plant, and larger diameter stems and branches. An increase in defensive chemicals may also result under such conditions (van Asch and Visser, 2007). Either of these changes to host physiology would influence palatability to insects, though the impacts on pests differ by species. For example, under increased CO_2 levels the winter moth (*Operophtera brumata*) consumes more oak (*Quercus robur*) leaves due to a reduction in leaf toughness, while the gypsy moth (*Lymantria dispar*) exhibits normal pupation weight but requires a longer time to develop as a result of an increase in tannin concentrations (van Asch and Visser, 2007).

Nitrogen Deposition

Anthropogenic emissions of nitrogenous air pollutants and their subsequent deposition are part of the larger phenomenon of global changes and can also have impacts on forest health. Increased nitrogen levels in the leaves of trees and plants can result in an immediate increase in the incidence of a number of pathogens (Burdon, Thrall and Ericson, 2006). For example, infections caused by the common pathogen, *Valdensia heterodoxa,* on *Vaccinium myrtillus,* a dominant understorey plant in boreal coniferous forests, are boosted by increased nitrogen availability resulting in premature leaf-shedding (Burdon, Thrall and Ericson, 2006). This defoliation subsequently promoted a shift from *V. myrtillus* dominance to grass dominance thereby affecting community structure.

Extreme Events

Besides drought, climate change may affect the frequency and intensity of other extreme climate-related events, with subsequent impacts on forest health. Direct damage to trees or alterations in the ecosystem may increase their susceptibility to pest outbreaks. Windstorms and lightning strikes can damage trees and allow entry of pathogens and secondary insect pests as well as causing mechanical breakdown in normal physiological function.

INDIRECT IMPACTS ON COMMUNITY ECOLOGY

Climate change is expected to alter the relationships between pests, their environment and other species, such as natural enemies, competitors and mutualists, leading to changes in the structure and composition of natural communities. The observed and predicted changes on species abundance and in phenological patterns and distributions of individual species are likely to alter species interactions within communities. Since individual species will respond to climate change in different ways and at different temporal scales there is a good possibility that some highly evolved relationships will be impacted. Interactions that involve two or more trophic groups, such as plant-herbivore, plant-pollinator and host-parasitic interactions are likely to suffer the largest mismatch.

Phenological Changes

In a review of phenological changes of interacting species, Visser and Both (2005) noted that insects have advanced their phenology faster (early eggs hatching and early mi-gration return date) than their hosts (budburst and flowering). They have also advanced their period of peak abundance more than their predators (laying date and migration arrival of birds). For example, the disruption of synchrony between the winter moth (*Operophtera brumata*) hatching and bud burst of its host oak trees has in turn resulted in an asynchrony between the pest and one of its predators, the great tit (*Parus major*), which relies on the caterpillars to feed their young. Such climate-induced phenological changes are clearly resulting in a great deal of asynchrony between interacting species which will ultimately influence community structure, composition and diversity. Distributional changes and range shifts interfere with community relationships as expanding species will begin to interact with other species in new environments with which previous interaction may have been limited or non-existent. The altitudinal spread of the pine processionary caterpillar (*Thaumetopoea pityocampa*) in the Sierra Nevada mountains of southeastern Spain for example, has resulted in the pest encountering a new host tree, the endemic Scots pine (*Pinus sylvestris* var. *nevadensis*). Increased attacks by *T. pityocampa* could have deleterious effects on this endemic mountain tree species.

Species capable of responding to climate change by increasing their range will also benefit from the lack of competitors and natural enemies in their new environment. Species expansions may not be promptly followed by that of its enemies, as in the case of the pine processionary moth, and in some cases, the synchronization between host and enemy or parasitoid may not be maintained under new temperature conditions, as is the case with the winter moth.

Some pathogens may benefit from the improved survival and spread of their insect vectors. For example, the vectors of Dutch elm disease (*Ophiostoma novo-ulmi*), *Scolytus scolytus* and *S. multistriatus* may be more active during periods of elevated temperature which would ultimately result in increased spread of the fungus.

Insects and diseases have been noted to respond to warming in all the expected ways, from changes in phenology and distribution to influencing community dynamics and composition. While some impacts of climate change may be beneficial in terms of protecting forest health (*e.g.* increase winter mortality of some insect pests due to thin snow cover; slower larval development and increased mortality during droughts), many impacts will be quite detrimental (*e.g.* accelerate insect development rate; range expansions of pests). Forest communities already exist and survive across a wide range of climatic conditions, suggesting that forests will persist under altered climatic conditions. However, the transitions, when pests expand into new territories without the checks and balances provided by natural enemies, or encounter either a new host species or a large expanse of their natural host species, may create opportunities for significant episodes of outbreaks, of reductions in forest growth and of tree mortality. Predicting and managing these transitions is where the challenges lie.

CLIMATE CHANGE AND FOREST PEST SPECIES

Some examples of forest insect pests, diseases and other pests which have been impacted or are predicted to be impacted by climate change are presented below. Information on non-forest pests is also provided to enable a better understanding of the potential impacts of climate change on forest health.

INSECTS

Coleoptera

Agrilus pannonicus (Buprestidae): A number of Buprestid beetles of the genus *Agrilus* have been linked to oak decline. Incidences of these species have increased worldwide (both in their countries of origin and by international movement) and their impacts are being linked to host tree stress potentially caused by climate change. For example, *Agrilus pannonicus* (=*A. biguttatus* (Fabricius)) has recently been associated with a European oak decline throughout its natural range and has increased in incidence in several countries including France, Germany, Hungary, Poland and the Netherlands, and the UK where it is believed to be contributing to oak decline. Infestations can result in extensive tree mortality which, combined with other factors involved in the decline, can drastically alter the species composition of oak forests.

Dendroctonus frontalis Zimmermann (Scolytidae) - Southern pine beetle: *Dendroctonus frontalis* is considered to be one of the most damaging species of bark beetles in Central America and southern areas of North America. It is a major pest of pines and has a wide distribution occurring from Pennsylvania in the United States south to Mexico and Central America. Populations can build rapidly to outbreak proportions and large numbers of trees are killed. Initial attacks are generally on weakened trees however *D. frontalis* is capable of killing otherwise healthy trees. This beetle kills trees by a combination of two factors: girdling during construction of egg galleries; and the introduction of blue stain fungi of the genus *Ophiostoma.* Because of their short generation time, high dispersal abilities and broad distribution of suitable host trees, the southern pine beetle has the potential to respond quickly and dramatically to any changes in climate.

In October 1998, Hurricane Mitch hit Central America, causing floods and mudslides that ravaged local communities, forests and infrastructure. In the years that followed an unprecedented regionwide outbreak of pine bark beetles, mainly *D. frontalis* in association with other *Dendroctonus* and *Ips* species, destroyed over 100 000 ha of pine forest. As most of the standing dead and felled trees were left on site, fuel loads were drastically increased thereby resulting in extensive wildfires. With climate change expected to increase the frequency and severity of extreme events such as hurricanes, the potential for future devastating impacts on forests from both the initial disturbance and its cascading effects (*i.e.* other disturbances such as pest outbreaks and fire) is quite high.

Warmer temperatures attributed to climate change have also influenced the southern pine beetle resulting in range expansions in the United States. Laboratory measurements and published records of mortality in wild populations indicate that a temperature of -16 °C or less result in almost 100 percent mortality of the pest, thereby limiting its distribution in its current northern range. It was predicted that an increase in temperature of 3 °C would enable outbreaks to occur approximately 178 km farther north than in historical times (Ungerer, Ayres and Lombardero, 1999). Recent outbreaks of the southern pine beetle in northern and high-altitude ecosystems, where they were previously rare or absent, have been attributed to a warming trend of 3.3 °C in minimum winter air temperatures in the southeastern US from 1960-2004. This northern expansion is about as predicted by Ungerer, Ayres and Lombardero (1999).

The southern pine beetle has also possibly adapted ways to increase survival in cooler climes. Tran *et al.* (2007) showed, through field and laboratory studies of a northern population, that prepupae were more cold tolerant (by more than 3 °C) than pupae, adults and feeding larvae, and that the winter life stage structure was strongly biased toward this most cold-tolerant life stage. This tendency to overwinter in a cold tolerant life stage could be a coincidence however rather than a true adaptation.

Dendroctonus ponderosae (Hopkins) (Scolytidae) - Mountain pine beetle : The mountain pine beetle (*Dendroctonus ponderosae*) is the most destructive pest of mature pines in North America, particularly lodgepole pine (*Pinus contorta*). In the western United States, outbreaks have been increasing in area after several years of drought (Tkacz, Moody and Villa Castillo, 2007). A major epidemic of this pest has also been ongoing in western Canada (British Columbia (BC), and more recently, Alberta) for several years and even with large-scale efforts to mitigate the impacts of the pest, millions of trees have been killed. A record of over 10 million hectares of pines were recorded as infested during 2007 aerial overview surveys in BC, with 860 973 ha of this located in provincial parks and protected areas. It has been predicted that if the beetle continues to spread at its current rate as much as 80 per cent of mature pine in BC will be dead by 2013. The large numbers of dead and dying trees have also increased the risk of wildfires.

The problem has been exacerbated by successive years of mild winters, resulting in decreases of mortality of overwintering stages and generation time. Their life cycle is generally completed in one year; warmer temperatures can result in two generations per year while cooler ones may results in one generation every two years (Amman, McGregor and Dolph, 1990). Drought conditions associated with warmer temperatures have also weakened the trees and increased their susceptibility to the beetles. Warmer temperatures have thus opened up previously climatically unsuitable mature pine stands to the pest.

Dendroctonus rufipennis (Kirby) (Curculionidae) - Spruce beetle : *Dendroctonus rufipennis* is a North American pest of spruce, particularly white spruce (*Picea glauca*) and black spruce (*P. mariana*) in the north, Engelmann spruce (*P. engelmannii*) and sitka spruce (*P. sitchensis*) in the west, and red spruce (*P. rubens*) in the east. It tends to attack weakened or windthrown trees and outbreaks are mostly linked to predisposing factors. As a result it can be expected that the impacts of climate change on trees and forests could enhance spruce beetle outbreaks.

In fact, Hebertson and Jenkins (2008) investigated the impact of climate on spruce beetle outbreaks in Utah and Colorado, USA between 1905 and 1996 and found that historic outbreak years in the intermountain region were related to generally warm fall and winter temperatures and drought conditions. Similarly outbreaks in both Canada (Yukon Territory) and the US (Alaska) appear to be related to extremely high summer temperatures which influenced spruce beetle population size through a combination of increased overwinter survival, a halving of the maturation time from two years to one year, and regional drought-induced stress of mature host trees.

HEMIPTERA

Aphids (Aphididae)

With short generation times and low developmental threshold temperatures, aphids are a group of insects that can be expected to be strongly influenced by environmental and climatic changes. In general, it has been predicted that aphids will appear at least eight days earlier in the spring within 50 years, though the rate of advance will vary depending on location and species. This could potentially result in greater damage to host plants depending on the phenology of host plants and natural enemies.

Zhou *et al.* (1995), for example, investigated the timing of migration in Great Britain for five aphid species (*Brachycaudus helichrysi, Elatobium abietinum, Metopolophium dirhodum, Myzus persicae, Sitobion avenae*) over a period of almost 30 years and concluded that temperature, especially winter temperature, is the dominant factor affecting aphid phenology for all species. They found that a one degree Celsius increase in average winter temperature advanced the migration phenology by 4-19 days depending on species.

Elatobium abietinum (Walker) (Aphididae)

The green spruce aphid (*Elatobium abietinum*) is also believed likely to benefit from the increase in winter survival, leading to more intense and frequent defoliation of host spruce trees (*Picea* spp.).

This aphid is native to Europe but has also been reported in both North and South America. Infestations in Great Britain have resulted in large losses of spruce foliage and height both during the active infestation and in subsequent years.

Westgarth-Smith *et al.* (2007) showed that warm weather associated with a positive North Atlantic Oscillation (NAO) index caused spring migration of *E. abietinum* to start earlier, last longer and contain more aphids.

Positive NAO values correspond to warmer atmospheric conditions over Great Britain. Since global warming is believed to increase NAO variability, shifting the system to more positive values, this will most likely lead to further increases in aphid activity and more damage to spruce trees and forests in the area.

HYMENOPTERA

Cephalcia arvensis Panzer

The spruce webspinning sawfly is monophagous on spruce (*Picea*) and endemic to the spruce range in Eurasia, where outbreaks have been seldom recorded.

From 1985-1992 however there was a sudden outbreak of the sawfly in the Southern Alps during which populations developed an annual life cycle and grew exponentially, causing repeated defoliations resulting in extensive tree death. *Cephalcia* species generally show low fecundity and have an extended diapause of a few years that is stimulated by low temperatures at pupation time.

The outbreak corresponded to a period of high temperatures and low precipitation and severe water stress for the host trees. As a result, the insect was able to adapt to the new climate resulting in lower mortality, faster development and higher feeding rates of the sawfly. In addition, the sudden increase in population density was not quickly followed by that of natural enemies, thereby allowing for unlimited population growth.

Neodiprion sertifer (Geoffroy) (Diprionidae)

The European pine sawfly *Neodiprion sertifer* is an important pest species on pines in Europe, northern Asia, Japan and North America where it was introduced. It is one of the most serious defoliators of Scots pine (*Pinus sylvestris*) forests in northern Europe. Virtanen *et al.* (1996) suggested that outbreaks of the sawfly on Scots pine in eastern and northern Finland are prevented by low winter temperatures which kill eggs, and predicted that outbreaks would become more common with winter warming. A high variation in freezing avoidance of eggs was also noted which would allow *N. sertifer* to adapt to predicted climate change and spread its distribution northwards.

LEPIDOPTERA

Warmer temperatures have been linked to increasing populations of forest Lepidoptera species.

Butterflies

While only a few species of butterflies are considered to be serious forest pests, some of the best, and most researched, examples of the impacts of climate change on insect distributions and phenology have been butterflies. The geographic ranges of many species have shifted northward and upwards in elevation associated with climate warming, leading to increases in species richness at high latitudes and elevations and in some cases possible local extinction at lower altitudes.

Range expansions in butterflies have been well documented and changes in butterfly phenology have also been reported. In the UK, species have been advancing their flight periods by approximately 2-10 days for every 1 ºC increase in temperature (Roy and Sparks, 2000; Menéndez, 2007). Similar changes in phenology as a response to warming has been noted in Spain where butterflies have advanced their first appearance by 1-7 weeks in 15 years and in California, USA which has seen an advancement of approximately eight days per decade (Forister and Shapiro 2003).

Some species-specific examples of the influence of climate change include the following.

- The African Monarch butterfly (*Danaus chrysippus*) has spread northward, establishing its first population in southern Spain in 1980 followed by the establishment of multiple populations along the east coast of Spain.
- Warmer temperatures have increased survival and facilitated a latitudinal and altitudinal range expansion of *Atalopedes campestris* in the western USA.
- Edith's checkerspot butterfly (*Euphydryas editha*) has shifted its distribution northwards and also upwards in altitude in North America. Populations at the northern edge of the species range in

Canada and also at higher altitudes within the main range have experienced increased survival whereas populations at the southern edge in Mexico have declined.

- In Europe the speckled wood butterfly (*Parage aegeriae*) has increased its range northward beyond its original, primary host (Logan, Régnière and Powell, 2003).
- The black-veined white butterfly, *Aporia crataegi,* has expanded its altitudinal range in the mountains of the Sierra de Guadarrama of central Spain resulting in local population extinctions at lower warmer altitudes. While climate is becoming less of a limiting factor in its distribution at higher altitudes, it is however limited by the absence of host plants.

Choristoneura fumiferana

The spruce budworm, *Choristoneura fumiferana,* is a major defoliator of coniferous forests across North America. Balsam fir (*Abies balsamea*) is the preferred host but they readily attack white, red and black spruce (*Picea glauca, P. rubens, P. mariana* respectively) and may even be feed on tamarack (*Larix* spp.) and hemlock (*Tsuga* spp.). Outbreaks of this budworm can persist for 5-15 years with periods of 20-60 years in between. In eastern Canada the period of population cycle has averaged 35 years over the last 270 years. During uncontrolled outbreaks they can kill almost all trees in dense, mature stands of fir.

Climatic influences on life history traits are considered a major factor in restricting outbreaks and as a result, a changing climate is expected to impact the severity, frequency, and spatial distribution of spruce budworm outbreaks (Logan, Régnière and Powell, 2003). The success of the insects in establishing feeding sites in the spring depends on initial egg weights and synchrony of their development with that of buds of their hosts which is strongly influenced by climatic factors (Volney and Fleming, 2000, 2007). This synchronization is critical in initiating outbreaks and thus determining the intensity of damage. However the spruce budworm is able to tolerate some asynchrony between spring emergence and vegetative shoot development as the second instars have adapted morphologically and behaviourally allowing them to mine needles.

In parts of its range, particularly at northern extremes, temperature can also influence the duration of outbreaks as collapses are often associated with the loss of suitable foliage often as a result of late spring frosts. Normal collapse of the outbreak in the core range of host trees is associated with mortality caused by natural enemies late in the larval stage. Natural enemies of the spruce budworm, *C. fumiferana,* are less effective at higher temperatures and therefore climatic factors have the potential to enable massive outbreaks of this pest providing there is suitable availability of host trees.

Epirrita autumnata (Geometridae)

Epirrita autumnata is a holarctic species that has been expanding its outbreak range in some areas. In the Nordic countries of Europe, *Epirrita autumnata* outbreak cycles are typically most prevalent in northernmost and continental birch forests but during the past 15-20 years it has expanded into the coldest, most continental areas previously protected by extreme winter temperatures. This pest overwinters in the egg stage and therefore the level of egg survival is dictated by minimum winter temperatures. Virtanen, Neuvonen and Nikula (1998) investigated the relationship between *E. autumnata* egg survival and minimum winter temperatures in northernmost Finland. They predicted that climate warming would result in a two-third reduction of the area of forests with winter temperatures cold enough to keep *E. autumnata* populations low by the middle of the next century. A rise in winter temperatures therefore will likely increase the area of forest susceptible to damage by the autumnal moth.

Lymantria dispar Linnaeus

The gypsy moth, *Lymantria dispar*, is a significant defoliator of a wide range of broadleaf and even conifer trees. While low population levels can exist for many years without causing significant damage, severe outbreaks can occur resulting in severe defoliation, growth loss, dieback and sometimes tree mortality. Two strains of gypsy moth exist – the Asian strain, of which the female is capable of flight; and the European strain, of which the female is flightless.

The Asian strain is native to southern Europe, northern Africa, central and southern Asia, and Japan and has been introduced into Germany and other European countries where it readily hybridizes with the European strain. It has also been introduced but has not established in Canada, the US and the UK (London). The European strain is found in temperate forests throughout Western Europe and has been introduced into Canada and the US. The gypsy moth is considered a significant pest in both its native and introduced ranges.

There has also been a noted increase in outbreaks in areas previously unaffected by this pest such as the Channel Islands (Jersey) and new areas in the UK (Aylesbury, Buckinghamshire). In Canada the spread of the gypsy moth has so far been prevented by climatic barriers and host plant availability as well as by aggressive eradication of incipient populations (Régnière, Nealis and Porter, 2008). However it is predicted that the gypsy moth will be able to extend its range in North America as a result of higher overwinter survival of egg stages because of milder winters and higher accumulation of day degrees for larval development. Similar predictions have been made for other areas. For example, Pitt, Régnière and Worner (2007) noted an increase in the probability of establishment of the gypsy moth in New Zealand, particularly in the South Island.

Increasing atmospheric concentrations of CO_2 may also influence the severity of gypsy moth outbreaks. Larval performance on host plants grown under elevated CO_2 varies depending on the species, being reduced on some hosts such as aspen and increased on others, such as oak.

Lymantria monacha

Lymantria monacha is a major pest of broadleaved and coniferous trees in Europe and Asia. Defoliation by nun moth larvae can kill host trees especially conifers and has caused extensive losses despite intervention with biological and chemical insecticides. In parts of Europe, the occurrence of outbreaks has increased possibly as a result of the establishment of extensive pine plantations in poor quality areas or as a result of a changing climate. It has been predicted that nun moth will spread northwards in Europe because of higher accumulated day degrees and improved overwinter survival. Using modelling software, Vanhanen *et al.* (2007) predicted that climate warming would shift the northern boundary of distribution north by approximately 500-700 km and the southern edge of the range would retract northwards by 100-900 km.

Operophtera brumata Linnaeus

Operophtera brumata is distributed throughout Europe, North Africa, Japan and Siberia and has also been introduced into Canada and the US. It feeds on a variety of deciduous trees and shrubs including apricot, cherry, apple, plum, blueberry, crab-apple, sweet chestnut, red currant and black currant, oaks, maples, basswood and white elm.

Climate change is impacting the spread of the winter moth. In the Nordic countries of Europe, Jepsen *et al.* (2008) noted that *O. brumata* had been climatically restricted to more southern and near-coastal locations in the regions but warmer temperatures has resulted in expansions in its outbreak area further northeast. While increased temperatures appear to assist the winter moth expand its distribution, it appears that they do not have the same impact on its natural enemies which may allow populations of this pest to grow unchecked.

Climate change has affected the phenology of many species in different ways. In the Netherlands over the past 25 years, early spring temperatures have increased while winter temperatures have not. As a result a climate change induced asynchrony has occurred between winter moths and their host, pedunculate oak, *Quercus robur*, with eggs hatching before bud burst (van Asch and Visser, 2007). Such a situation leaves no food for the larvae resulting in starvation and death. This also has implications for other species that depend on the larvae for food such as the great tit (*Parus major*) which feed *O. brumata* caterpillars to their young. While both egg hatch and bud burst have advanced over the last 25 years, egg hatch has advanced much more leading to a decrease in synchrony from a few days to almost 2 weeks. However others have noted that, while warmer temperatures have lead to earlier egg hatch, the autumnal pupal diapause of the winter moth is prolonged at higher temperatures thereby counteracting the impact and resulting in a life cycle that is not shortened overall. Differences between observations of synchronicity between moth egg hatch and host bud burst may result from regional, intra-specific differences.

Thaumetopoea pityocampa **(Thaumetopoeidae)**

The pine processionary caterpillar, *Thaumetopoea pityocampa*, is considered one of the most important pests of pine forests in the Mediterranean region. It is a tent-making oligophagous caterpillar that feeds gregariously and defoliates various species of pine and cedar. The life cycle of the pine processionary caterpillar is typically annual but may extend over two years at high altitudes or in northern latitudes. At northern latitudes and at higher altitudes, adults emerge earlier.

Climate change is having clear impacts on the distribution of this important forest pest. Battisti *et al.* (2005) reported a latitudinal expansion in north-central France of 87 km northwards from 1972-2004 and an altitudinal shift of 110-230 m upwards in the Alps of northern Italy from 1975-2004 and attributed the expansions to reduced frequency of late frosts, which increases survival of overwintering larvae, as a result of a warming trend over the past three decades. In the last ten years the pine processionary caterpillar has spread almost 56 km northward in France.

During the summer of 2003, the warmest summer in Europe in the last 500 years, *T. pityocampa* exhibited an unprecedented expansion to high elevation pine stands in the Italian Alps, increasing its altitudinal range limit by one third of the total altitudinal expansion over the previous three decades. This unusual and fast spread has been attributed to increased nocturnal dispersal of females during the unusually warm night temperatures.

The gradual warming of the region has allowed the pest to maintain its presence at this altitude because of increased larval survival. In the Sierra Nevada mountains of southeastern Spain, *T. pityocampa* has expanded to higher elevations over the last 20 years as a result of increasing mean temperature. Relict populations of Scots pine (*Pinus sylvestris* var. *nevadensis*) occurring within this newly expanded range of the caterpillar are being increasingly attacked, particularly in warmer years. This range expansion caused by climate change has potentially devastating consequences for this endemic mountain species which is likely to suffer from the direct effects of climate change as well.

Given that the present distribution of the *T. pityocampa* is not constrained by the distribution of its hosts, that warmer winters will increase winter larval feeding activity, that the probability of lower lethal temperature will decrease, it can be expected that the improved survival and spread into previously hostile environments will continue.

Thaumetopoea processionea

Native to central and southern Europe, *Thaumetopoea processionea* is a major defoliating pest of oak. Since the late 20th century it has been expanding its range northwards and is now firmly established in Belgium, Denmark, northern France, and the Netherlands and has been reported from southern Sweden and the UK. It is believed that the northward progression of the oak processionary moth is due to improved synchrony of egg hatch and reduction of late frosts as a result of warmer temperatures.

***Zeiraphera diniana* Guenée**

The larch bud moth, *Zeiraphera diniana,* is a European pest that has been defoliating large areas of larch forests in the Alps every 8-10 years for centuries. It has an annual life cycle, overwintering as an egg on the larch branches and feeding on the needles as soon as the bud breaks. As such synchrony between egg hatch and bud burst is critical. Increased temperatures associated with climate change have affected this relationship leading to asynchrony and reduced incidences of the moth in Switzerland. It has been reported that abnormally high temperatures result in unusually high egg mortality.

PATHOGENS

ASCOMYCOTA

***Mycosphaerella pini* Rostrup**

Mycosphaerella pini is a fungus that infects and kills the needles of *Pinus* spp. resulting in significant defoliation, stunted growth and eventually death of host trees although susceptibility among pine species does vary. Native to temperate forests of the northern hemisphere, it is a major pest of pine plantations in the Southern Hemisphere, where both the host and the pathogen have been introduced. The widely planted *P. radiata* is particularly susceptible and many forests planted with this species in the Southern Hemisphere, particularly in East Africa, New Zealand and Chile, have been devastated by this needle blight. This pathogen has forced managers in some areas to abandon the planting of *P. radiata* and depend more on other tree species.

In its native range *M. pini* normally causes little damage, but since the late 1990s it has been causing extensive defoliation and mortality in young plantations of lodgepole pine (*Pinus contorta* var. *latifolia*) in northwestern British Columbia, Canada. Mortality of mature lodgepole pines has been observed in mixed-species stands, where scattered pine represents only a small proportion of stand composition; this represents a globally unprecedented occurrence for *M. pini*. The current epidemic coincides with a prolonged period of increased frequency of warm rain events throughout the mid-to-late 1990s allowing for the rapid spread and increased rates of infection. Unlike many other pests, changes in precipitation patterns may be more important than changes in temperature for predicting the spread and impact of *M. pini*.

BASIDIOMYCOTA

Armillaria species (Agaricales: Marasmiaceae)

Armillaria species are common worldwide pathogens of trees, woody shrubs and herbaceous plants that can cause wood decay, growth reduction and even mortality, particularly in trees stressed by other factors, or in young trees planted on sites from which infected hosts have been removed. *Armillaria* species can become more aggressive and damaging when elevated temperatures cause drought stress thereby reducing tree defences. Tree physiological condition in general may be an important factor in controlling the impacts of *Armillaria* species, and climate change may affect their epidemiology.

OOMYCOTA

Phytophthora cinnamomi

Phytophthora cinnamomi is considered one of the most widely distributed and destructive forest pathogens. It has a wide host range infesting over 1000 species resulting in root rot and cankering. The native range is unknown but it is believed to be from Southeast Asia and southern Africa. Currently the pathogen can be found in most temperate and subtropical areas in the world in Africa, Asia and the Pacific, Europe, Latin America and the Caribbean, Near East, and North America. In most countries it is only known in nurseries but in Europe (France, Italy, Spain, Portugal) it has observed in natural environments.

Temperature, moisture and pH all influence the growth and reproduction of the fungus. In a study on the impacts of climate warming on *P. cinnamomi*, Bergot *et al.* (2004) predicted a potential range expansion of the disease in Europe of one to a few hundred kilometres eastward from the Atlantic coast within one century.

Phytophthora ramorum Werres

Phytophthora ramorum causes a very serious disease called sudden oak death which causes extensive mortality of tanoak and oaks. It is also associated with disease on ornamental plants and other broadleaf and conifer trees. This pathogen is a significant problem in both North American and European forests and nurseries. The geographic origin of *P. ramorum* is unknown; it is believed that it has been introduced independently to Europe and North America from an unidentified third country.

The pathogen likely disperses through a variety of means. Sporangia may be dispersed locally by rain splash, wind-driven rain, irrigation or ground water, soil and soil litter. Bark and ambrosia beetles are commonly found on infected trees but their potential role of vectors has not yet been investigated. Consequently, changes in climate, precipitation and temperature in particular, will likely produce more optimal conditions for the pathogen resulting in an increase in disease occurrence.

OTHER PESTS

NEMATODES

In general there is a close correlation between soil temperatures and the distributions of some plant-parasitic species of nematode. For example, *Meloidogyne incognita*, previously deemed limited to the Mediterranean area, was recently found in the Netherlands. It is also believed that a one degree Celsius rise in temperature would allow *Longidorus caespiticola* to become established further north in Great Britain.

Bursaphelenchus xylophilus

The pine wilt nematode, *Bursaphelenchus xylophilus*, is the causal agent of pine wilt disease and is spread by *Monochamus* beetles. Native to North America where it is not considered a serious pest, the nematode is a major threat to Asian and European pine forests and has resulted in extensive tree mortality in countries where it has been introduced.

Changes in both temperature and precipitation are likely to impact the spread of the nematode and the severity of damage caused by the disease. Pine wilt disease is most prevalent in warm climates as the nematode completes its life cycle in 12, 6 and 3 days at 15, 20 and 30°C, respectively. High temperatures and low precipitation in summer cause accelerated damage through their impacts on vector activity, propagation of the nematode and water stress on trees. In Japan, while annual tree losses to the disease have gradually decreased, infestations have spread into northern areas and into forests at higher elevations as a result of increased temperatures.

The evidence presented from this desk review shows that climate change is having considerable and widespread impacts on forest health worldwide, and, as a result, on the forest sector. Clearly, if such climatic and ecological changes are now being detected when the globe has warmed by an estimated average of only 0.74 °C, it can be expected that many more impacts on species and ecosystems will occur in response to changes in temperature to levels predicted by IPCC. Conversely there are some indications that the interrelated effects of climate on tree hosts and the direct influence on natural enemies may make the overall effect difficult to predict and it is considered by some that not all climate change scenarios will be detrimental. The challenge to understanding climate change effects is not just in obtaining information on the impacts of temperature, precipitation and other climatic factors on forests and pests but also acquiring knowledge on the interaction between the different climate change factors, and how climate change impacts disturbances and vice versa. While a fair amount of information is already available concerning the impacts of climate change on the world's species and ecosystems, much more is needed.

From the perspective of forests, considerably more information is needed on the impacts on forests, forest pests and the complex relationships relating to climate change. Much of the information available comes from Europe and North America so there is a clear need for increased research in other regions. The most commonly studied insects belong to the orders containing butterflies, moths and aphids while there is only limited information on coleopterans and scant information on the effect of climate change on symbionts and host dynamics. Further detailed studies of important forest pests would allow for the development of pest management strategies for the future and assist forest managers and policy-makers to better prepare for the challenge of dealing with climate change.

The forest sector needs effective monitoring and detection activities to allow for quick action in the face of changing or increasing pest outbreaks including continual pest risk assessments. There is also a need for alternative practices to reduce subsequent vulnerability of forests, such as planting pest tolerant trees identified through breeding programmes; noting however that it is unlikely that such programmes can predict new pest risks in a timely fashion due to shifting species adapting to new environments. Comprehensive risk assessments as well as enhanced knowledge management systems using a variety of information technologies such as simulation models, geographic information systems (GIS) and remote sensing could also play a role in protecting forest health from the impacts of climate change and forest pests

9

Climate Change in Modern World

CLIMATE SYSTEM

The key to understanding global climate change is to first understand what global climate is, and how it operates. At the planetary scale, the global climate is regulated by how much energy the Earth receives from the Sun. However, the global climate is also affected by other flows of energy which take place within the climate system itself. This global climate system is made up of the atmosphere, the oceans, the ice sheets (cryosphere), living organisms (biosphere) and the soils, sediments and rocks (geosphere), which all affect, to a greater or less extent, the movement of heat around the Earth's surface.

The atmosphere plays a crucial role in the regulation of Earth's climate. The atmosphere is a mixture of different gases and aerosols (suspended liquid and solid particles) collectively known as air. Air consists mostly of nitrogen (78%) and oxygen (21%). However, despite their relative scarcity, the so-called greenhouse gases, including carbon dioxide and methane, have a dramatic effect on the amount of energy that is stored within the atmosphere, and consequently the Earth's climate.

These greenhouse gases trap heat within the lower atmosphere that is trying to escape to space, and in doing so, make the surface of the Earth hotter. This heat trapping is called the natural greenhouse effect, and keeps the Earth 33°C warmer than it would otherwise be. In the last 200 years, man-made emissions of greenhouse gases have enhanced the natural greenhouse effect, which may be causing global warming.

The atmosphere however, does not operate as an isolated system. Flows of energy take place between the atmosphere and the other parts of the climate system, most significantly the world's oceans. For example, ocean currents move heat from warm equatorial latitudes to colder polar latitudes. Heat is also transferred via moisture. Water evaporating from the surface of the oceans stores heat which is subsequently released when the vapour condenses to form clouds and rain. The significance of the oceans is that they store a much greater quantity of heat than the atmosphere. The top 200 metres of the world's oceans store 30 times as much heat as the atmosphere. Therefore, flows of energy between the oceans and the atmosphere can have dramatic effects on the global climate.

The world's ice sheets, glaciers and sea ice, collectively known as the cryosphere, have a significant impact on the Earth's climate. The cryosphere includes Antarctica, the Arctic Ocean, Greenland, Northern Canada, Northern Siberia and most of the high mountain ranges throughout the world, where sub-zero temperatures persist throughout the year. Snow and ice, being white, reflect a lot of sunlight, instead of absorbing it. Without the cryosphere, more energy would be absorbed at the Earth's surface rather than reflected, and consequently the temperature of the atmosphere would be much higher.

All land plants make food from the photosynthesis of carbon dioxide and water in the presence of sunlight. Through this utilisation of carbon dioxide in the atmosphere, plants have the ability to regulate the global climate. In the oceans, microscopic plankton utilise carbon dioxide dissolved in seawater for photosynthesis and the manufacture of their tiny carbonate shells. The oceans replace the utilised carbon dioxide by "sucking" down the gas from the atmosphere. When the plankton die, their carbonate shells sink to the seafloor, effectively locking away the carbon dioxide from the atmosphere. Such a "biological pump" reduces by at least four-fold the atmospheric concentration of carbon dioxide, significantly weakening the Earth's natural greenhouse effect, and reducing the Earth's surface temperature.

COMPONENTS OF THE CLIMATE SYSTEM

OCEANS

The atmosphere does not respond as an isolated system. Like the atmosphere, the thermodynamic state of the oceans is determined by the transfer of heat, momentum and moisture to and from the atmosphere. Ignoring for the moment the other components of the climate system, these fluxes within this coupled ocean-atmosphere system exist in equilibrium.

Momentum is transferred to the oceans by surface winds, mobilising the global surface ocean currents. Surface ocean currents assist in the latitudinal transfer of sensible heat in a similar fashion to the process occurring in the atmosphere. Warm water moves poleward whilst cold water returns towards the equator. Energy is also transferred via moisture. Water evaporating from the surface of the oceans stores latent heat which is subsequently released when the vapour condenses to form clouds and precipitation.

The significance of the ocean is that it stores a much greater quantity of energy than the atmosphere. This is on account of both its larger heat capacity (4.2 times that of the atmosphere) and its much greater density (1000 times that of air). The vertical structure of the ocean Figure can be divided into two layers which differ in the scale of their interaction with the overlying atmosphere. The lower layer comprises the cold deep water sphere, making up 80% of the oceans' volume.

The upper layer, which has closest contact with the atmosphere, is the seasonal boundary layer, a mixed water sphere extending down only 100m in the tropics but several kilometres in polar regions. The seasonal boundary layer alone stores approximately 30 times as much heat as the atmosphere. Thus for a given change in heat content of the ocean-atmosphere system, the temperature change in the atmosphere will be around 30 times greater than that in the ocean. Clearly then, small changes to the energy content of the oceans could have considerable effects on global climate.

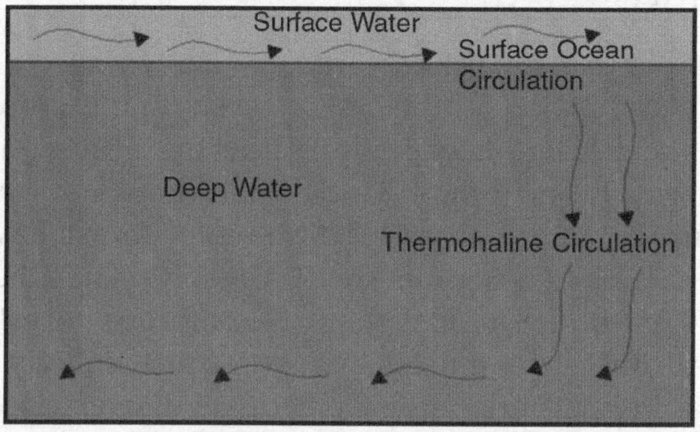

Fig. Vertical Structure and Circulation of the Oceans

Energy exchanges also occur vertically within the oceans, between the mixed boundary layer and the deep water sphere. Sea salt remains in the water during the formation of sea ice in the polar regions, with the effect of increased salinity of the ocean. This cold, saline water is particularly dense and sinks, transporting with it a considerable quantity of energy. To maintain the equilibrium of water (mass) fluxes, a global thermohaline + circulation exists, which plays an important role in the regulation of the global climate.

THE CRYOSPHERE

The cryosphere consists of those regions of the globe, both land and sea, covered by snow and ice. These include Antarctica, the Arctic Ocean, Greenland, Northern Canada, Northern Siberia and most of the high mountain ranges throughout the world, where sub-zero temperatures persist throughout the year. The cryosphere plays another important role in the regulation of the global climate system.

Snow and ice have a high albedo + (reflectivity), that is they reflect much of the solar radiation they receive. Some parts of the Antarctic reflect as much as 90% of the incoming solar radiation, compared to a global average of 31%. Without the cryosphere, the global albedo would be considerably lower. More energy would be absorbed at the Earth's surface rather than reflected, and consequently the temperature of the atmosphere would be higher. Indeed, during the Cretaceous Period (120 to 65 million years ago) evidence suggests there was little or no snow and ice cover, even at the poles, and global temperatures were at least 8 to 10°C warmer than today. The cryosphere also acts to decouple the atmosphere and oceans, reducing the transfer of moisture and momentum, so stabilising the energy transfers within the atmosphere. The formation of sea ice in polar regions can initiate global thermohaline circulation patterns in the oceans, which greatly influence the global climate system. Finally, the presence of the cryosphere itself markedly affects the volume of the oceans and global sea levels, changes to which can affect the energy budget of the climate system.

THE BIOSPHERE

Life may be found in almost any environment existing on Earth. Nevertheless, in a discussion on the climate system, it is convenient to regard the biosphere as a discrete component, like the atmosphere, oceans and cryosphere. The biosphere, both on land and in the oceans, affects the albedo of the Earth's surface. Large areas of continental forest have relatively low albedos compared to barren regions such as deserts. The albedo of deciduous forests is about 0.15 to 0.18 whilst that of coniferous forests is 0.09 to 0.15. Tropical rainforest reflects even less energy, approximately 7 to 15% of that which it receives. In comparison, the albedo of a sandy desert is about 0.3. Clearly, the presence of the continental forests affect the energy budget of the climate system.

The biosphere also influences the fluxes of certain greenhouse gases such as carbon dioxide and methane. Plankton in the surface oceans utilise the dissolved carbon dioxide for photosynthesis. This establishes a flux of carbon dioxide, with the oceans effectively "sucking" down the gas from the atmosphere. On death, the plankton sink, transporting the carbon dioxide to the deep ocean. Such primary productivity reduces by at least four-fold the atmospheric concentration of carbon dioxide, weakening significantly the Earth's natural greenhouse effect.

The biosphere also influences the amount of aerosols in the atmosphere. Millions of spores, viruses, bacteria, pollen and other minute organic species are transported into the atmosphere by winds, where they can scatter incoming solar radiation, and so influence the global energy budget. Primary productivity in the oceans results in the emission of compounds known as dimethyl sulphides (DMSs). In the atmosphere these compounds oxidise to form sulphate aerosols called marine non-sea-salt (nss) sulphate. These nss sulphates act as condensation nuclei for water vapour in the atmosphere, thus allowing the formation of clouds. Clouds have a highly complex effect on the energy budget of the climate system. Thus changes in primary productivity in the oceans can affect, indirectly, the global climate system. There are, of course, many other mechanisms and processes which couple the biosphere with the rest of the climate system, but the discussion has illustrated the major influences of the biosphere upon the global climate system.

THE GEOSPHERE

The fifth and final component of the global climate system is the geosphere, consisting of the soils, the sediments and rocks of the Earth's land masses, the continental and oceanic crust and ultimately the interior of the Earth itself. These parts of the geosphere each play a role in the regulation and variation of global climate, to a greater or lesser extent, over varying time scales.

Variations in global climate over tens of millions or even hundreds of millions of years are due to modulations within the interior of the Earth. Changes in the shape of ocean basins and the size of continental mountain chains (driven by plate tectonic processes) may influence the energy transfers within and between the coupled components of the climate system.

On much shorter time scales physical and chemical processes affect certain characteristics of the soil, such as moisture availability and water run-off, and the fluxes of greenhouse gases and aerosols into the atmosphere and oceans. Volcanism, although driven by the slow movement of the tectonic plates, occurs regularly on much shorter timescales. Volcanic eruptions replenish the carbon dioxide in the atmosphere, removed by the biosphere, and emit considerable quantities of dust and aerosols. Volcanic activity can therefore affect the energy budget and regulation of the global climate system.

TIME SCALE OF CLIMATIC CHANGE

The importance of considering different time scales when investigating climate change has already been identified. Climate varies on all time scales, in response to random and periodic forcing factors. Across all time periods from a few years to hundreds of millions of years there is a white (background) noise of random variations of the climate, caused by internal processes and associated feedback mechanisms, often referred to as stochastic or random mechanisms. Such randomness accounts for much of the climate variation, and owes its existence to the complex and chaotic behaviour of the climate system in responding to forcing. An essential corollary of the existence of random processes is that a large proportion of climate variation cannot be predicted.

Of far more relevance are the periodic forcing factors, for by understanding their mechanisms and the impacts they have on the global climate, it is possible to predict future climate change. How the climate system responds to periodic forcing factors, however, is often not clear. If it is assumed that the climate system responds in a linear fashion to periodic forcing, variations in climate will exhibit similar periodicity. If, however, the response of the system to forcing is strongly non-linear, the periodicities in the response will not necessarily be identical to the periodicities in the forcing factor(s). Frequently, the climate responds in a fashion intermediate between the two.

There are many climate forcing factors spanning an enormous range of periodicities. The longest, 200 to 500 million years, involves the passage of our Solar System through the galaxy, and the variations in galactic dust. These may be considered to be external forcing mechanisms. Other long time scale variations (10^6 to 10^8 years) include the non-radiative forcing mechanisms, such as continental drift, orogeny (mountain building) and isostasy (vertical movements in the Earth's crust affecting sea level).

The response of the climate system to this combination of forcing factors itself depends upon the different response times of the various components of the system. The overall climatic response will then be determined by the interactions between the components. The atmosphere, surface snow and ice, and surface vegetation typically respond to climatic forcing over a period of hours to days. The surface ocean has a response time measured in years, whilst the deep ocean and mountain glaciers vary only over a period spanning hundreds of years. Large ice sheets advance and withdraw over thousands of years whilst parts of the geosphere (*e.g.* continental weathering of rocks) respond only to forcing periods lasting hundreds of thousands to millions of years.

The response of the climate system to episodes of forcing can be viewed as a form of resonance. When the time period of forcing matches most closely the response time of a particular system component, the climatic response will be greatest within that component.

Milankovitch forcing, for example, with periods of tens of thousands of years will be manifest in the response of the ice sheets, and the overall response of the climate system will be dominated by changes within the cryosphere. In addition, longer response times of certain components of the climate system modulate, through feedback processes, the short term responses. The response of the deep ocean to short term forcing (*e.g.* enhanced greenhouse effect, solar variations, for example, will tend to attenuate or smooth the response of the atmosphere.

Throughout the remainder, it should be recognised that a range of time scales applies to climate forcing mechanisms, radiative and non-radiative, external and internal, and to the response of the different components of the climate system.

CLIMATE SENSITIVITY

The concept of feedback is related to the climate sensitivity or climate stability. It is useful to have a measure of the strength of various feedback processes which determine the ultimate response of the climate system to any change in radiative forcing. In general terms, an initial change in temperature due to a change in radiative forcing, $\Delta T_{forcing}$, is modified by the complex combination of feedback processes such that:

$$\Delta T_{final} = \Delta T_{forcing} + \Delta T_{feedback}$$

where $\Delta T_{feedback}$ is the temperature change resulting from feedback and ΔT_{final} is the overall change in temperature between the initial and final equilibrium states. The degree to which feedback processes influence the final climatic response is a measure of the sensitivity of the climate system.

Equation can be rewritten as follows:

$$\Delta T_{final} = f\Delta T_{forcing}$$

where f is called the feedback factor. When only one feedback mechanism is operative the solution to equation 6 is simple, assuming both f and $\Delta T_{forcing}$ are known.

When more than one feedback is operative, matters become more complicated. For two feedbacks, the net effect is given by:

$$f = f_1 f2/ (f_1 + f2 - f_1 f2)$$

where f_1 and f2 are the feedback factors of the two feedback processes. Clearly, the feedback factors are neither additive nor multiplicative. A feedback operating alone with a factor of 2 would double the initial climatic response to forcing. If a second feedback with factor 1.5 acted with it, the overall feedback would be enhanced by a factor of 6. It can be seen, then, that a combination of feedback processes could dramatically affect the climate as it responds to only a small change in radiative forcing.

The climate's sensitivity can be mathematically determined in another way. From satellite observations, it has been shown that changes in global temperature are approximately proportional to changes in radiative forcing. If we assume an instantaneous ‡ change in climate, from one equilibrium state to another, then:

$$\Delta Q = \lambda \Delta T$$

where ΔQ is change in radiative forcing (expressed in terms of the net downward radiative flux at the top of the troposphere), ΔT is the change in global temperature, and λ is a measure of the climate sensitivity.

Based on equation, the climate sensitivity is usually expressed in terms of the temperature change associated with a specified change in radiative forcing, usually a doubling of the atmospheric carbon dioxide content.

Thus, the carbon dioxide doubling temperature, ΔT_{2x}, is given by,

$$\Delta T_{2x} = \Delta Q_{2x}/\lambda$$

where ΔQ_{2x} is 4.2 Wm^{-2}. The magnitude (and sign) of ΔT_{2x} will depend on λ, the climate sensitivity, which is determined by the net affect of the climate feedback processes. Despite extensive climate modelling over the last 2 decades to understand the problem of contemporary global, it is this parametre that is proving hard to define numerically.

As explained earlier, the idea of static equilibrium and instantaneous climatic response represents an unrealistic situation.

To take into account the dynamic and transient nature of the climate's response to forcing, a more complicated equation linking ΔT and ΔQ is required if the evolution of the response with time is to be determined.

In the case:

$$\Delta Q = \lambda\Delta T + Cd\Delta T/\delta t$$

Here, the change in radiative forcing, DQ, is balanced by:

- The change in the outgoing radiative flux at the tropopause caused by the response of the climate system including feedback; and
- The energy stored in the system, $C\delta\Delta T/\delta t$, where C is the system's heat capacity and t is time.

The latter term in equation simulates the time-dependent nature of the climate system's response. The main contributor to the system's heat capacity is the world's oceans. The heat capacity of water is large compared to that of air and is therefore able to store much more energy. Additionally, the high heat capacity means that the oceans take time to heat up (or cool down) and hence slow the surface temperature response to any change in radiative forcing: the transient response will always be less than the equilibrium response.

Solution of Equation leads to the definition of the response time of the climate system, τ, such that:

$$\tau = C/\lambda$$

If the heat capacity of the climate system is large, the response time is large. Equally, if the climate sensitivity is small the response time is large. λ is also known as the radiative damping coefficient. Here, analogy is made between the response of the climate system and an oscillating spring. If a spring has a high damping coefficient, it will stop oscillating soon after it has been set in motion. Similarly, if the radiative damping coefficient is large, the climate will respond quickly and τ will be small.

CLOUD FORMATION AND CLIMATE

WATER AND HEAT STORAGE

How does water react to heating or cooling? How much energy can it store? It turns out that water is much more efficient in storing heat than any other common substance on Earth. To increase the temperature of fluid water by one degree Centigrade takes one calorie of heat energy for each gram of water. For comparison, to do the same for a rock takes 0.2 calories per gram, and for petroleum, 0.5 calories per gram. To change a gram of ice into water (without increasing the temperature) takes 80 calories. Conversely, when water freezes, it releases this same amount of heat into the environment. To change a gram of water into vapour requires 580 calories! This is what makes evaporation the best cooling mechanism around. Our body takes advantage of this fact when cooling itself by sweating: the evaporation of the sweat extracts heat from the surroundings, lowering the temperature of the skin. The heat expended in the phase change of water to vapour is not lost, but is contained in the vapour in latent form. When the vapour condenses, the heat is freed for warming the surrounding air.

The phase changes of water combined with its unique heat-related properties are intimately involved in all aspects of climate and weather. Water transfers and stores heat on an immense scale, and thereby evens out the temperature differences between day and night, summer and winter, tropics and polar areas.Consider the temperature differences between day and night: they are largest in the water-starved desert. By contrast, in areas where water is abundant evaporation during the day tends to lower the temperature, and condensation during the night (the familiar dew) tends to raise it. During the day, clouds protect the ground from the heat of the sun, and at night, from the black coldness of space. Seasonal differences are similarly tempered by the great heat reserves of water, hence the mildness of coastal climates. Likewise, geographic differences are moderated by the transfer of heat through water motion and water phase changes. Most of this transfer is by water vapour in the air. Warm winds blowing over warm ocean regions pick up the water, cooling the sea surface. In cold regions, the air becomes unable to hold the vapour. The vapour condenses and gives off its heat of evaporation. If it freezes to snow, it also gives off its heat of fusion, to the surrounding air. Thus, a blizzard brings warmth - a somewhat unexpected conclusion, perhaps, to those who have been surprised by a snowstorm during a hike in the mountains.

THE PROPERTIES OF CLOUDS

Clouds consist of fine water particles floating in air and which are dense enough to prevent the direct transmission of light. The particles can even be frozen, making them ice clouds. Condensation of water occurs when the moisture in the air exceeds the capability of the air to hold the water vapour (upon cooling moist air or mixing it with cold air) and when microscopic particles called aerosols are present on which the vapour can condense.

Aerosols are any small particle, solid or fluid, suspended in air. Abundance values typically range from 100 to 10,000 particles per cubic centimetre, for air over land, with higher values found in cities. Sizes vary greatly, but typically are near 0.1 micrometre or less. The particles originate from wind-blown sea-salt or dust, volcanic eruptions, from burning of vegetation, from combustion of coal and petroleum products, and other natural and anthropogenic processes. Aerosol particles serve as nuclei for condensation of water droplets and for growth of ice crystals and also influence the radiation balance directly. In the lower stratosphere, concentrations are extremely low, and much of the aerosols there consist of droplets of sulfuric acid.

Clouds come in many different types, depending on particle size and density, temperature and phase, thickness and elevation, cloud size and dynamics of change. The properties and the response of different types of clouds is crucial to the improvement of prediction of how the Earth's climate will respond to global warming. On the whole, higher temperature will produce greater moisture in the air, which will favour the formation of clouds in the cool regions of the atmosphere (which may move upward as the surface warms). Also, the general increase of pollution of the atmosphere that is associated with increased human activity (burning, agriculture, diesel engines) should favour the availability of cloud nuclei. Clouds also provide for "cloud Aledo," as they reflect much of the sunlight into space. Clouds also trap infrared radiation. How the balance between these two effects will shift is uncertain. From a geological viewpoint (considering the climate patterns on a warm Earth, before the presence of large glaciers) it seems reasonable to expect that increasing clouds in the tropics will cool the tropics and increasing clouds in high latitudes will warm the cold regions.

OCEAN CIRCULATION AND CLIMATE

When talking about the greenhouse effect, we have considered the fact that water on Earth occurs in all its three phases — gaseous, fluid and solid. As vapour in the air, water provides the strongest warming effect of all the greenhouse gases. As a fluid, it covers some seventy per cent of the Earth's surface, and if spread evenly over the planet, it would make a layer two miles thick. As a solid, it whitens the polar regions and the high latitudes in winter, and greatly increases the reflectivity of Earth's surface wherever it covers the ground. Ice also forms a protective layer on water wherever it covers the sea surface.

As we learned in the last lesson, the physical properties of water make it an ideal means for conditioning the climate on Earth. It stores heat better than most other substances – that's why tea takes some time to make, but then stays hot for quite a while. It takes up heat when moving from solid to fluid or from fluid to gas, and releases that heat again when moving through the phases in the opposite direction. The "latent" heat thus stored in water vapour is freed up with precipitation (and can then drive storms). This transfer of energy by latent heat is a hundred times more efficient than transfer by sensible heat, gram for gram.

Quite generally, the ocean surface is a source for heat in the atmosphere. The sea surface tends to stay at the same temperature for a long time, because of the high specific heat off water and the thickness of the mixed layer. Thus, the ocean surface stabilises the atmospheric flow patterns, which depend on heat supply. Without this stabilizing influence of the ocean, the weather would be much more fickle.

The general circulation of the atmosphere is closely related to the general circulation of the ocean. Winds blowing over the sea surface produce ocean currents. Winds also evaporate water, which precipitates elsewhere as rain. A large amount of heat is transferred to the atmosphere in this process. Evaporation and precipitation also affect the patterns of water density, which are inseparably tied to ocean circulation.

But how will the ocean's circulation respond to climate change? Although this sounds like a simple question, answering it is difficult. The ocean responds largely to winds, and wind fields respond to the distribution of heat, which depends on the ocean circulation. Within a logic loop like this, where the final answer depends on the initial guess, any small errors can be readily amplified into large errors in the prediction game.

STRAWBERRIES IN NORWAY

Differing Climates of Norway and Greenland

Ocean currents play an important role in transferring heat from the warm tropics to the cold polar regions. An example is the relatively mild climate of Norway that depends upon the warm waters brought north along the Norwegian coast through an extension of the Gulf Stream System. In Norway strawberries are called "earth-berries", and one can buy them from fruit stands along the road, when in season. We must remember that Oslo, the capital of Norway, is at 60°N, and that more than eighty per cent of Norway is north of Oslo, between 60 and 70°N.

By contrast, to the west across the Nordic Sea, there is Greenland: an island almost entirely covered by thick ice sheets. Its southern tip is also at 60°N here glaciers reach sea level and produce the icebergs which drift away into the North Atlantic threatening the safety of ships. There are no plants of any kind and certainly no strawberries. Halfway between Norway and southern Greenland lies Iceland with its impressive glaciers, some of which reach down to sea level. However, Iceland is quite green, with plenty of grazing for sheep (providing wool for the splendid Icelandic sweaters).

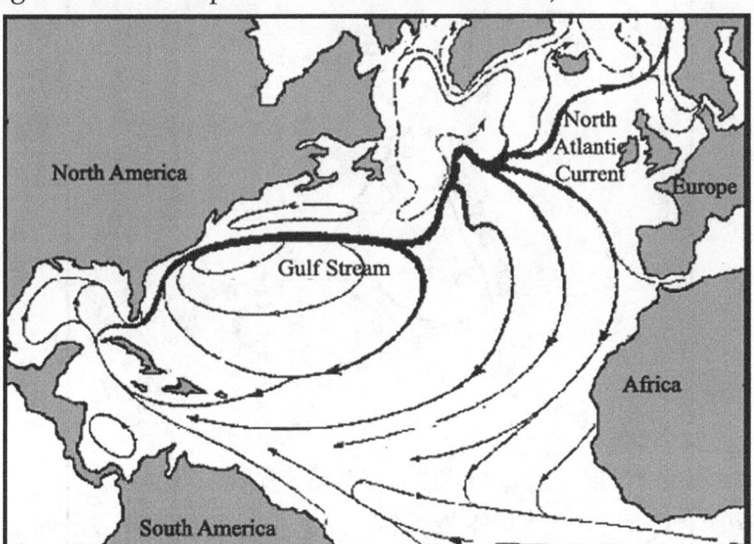

Map of the North Atlantic Ocean showing the Gulf Stream and the North Atlantic Current (Represented by the thick black lines). Cold currents originating in the north are represented by short-dashed lines. Note how some of the water gets recirculated into the North Atlantic subtropical gyre. Why the island covered with ice is called Greenland and the one that has pastures is called Iceland has to do with real estate speculation in Viking times, and need not concern us further. But we do have to ask, why are Norway and Iceland so entirely different from southern Greenland (which is, after all, situated in the same latitudinal belt of 60 to 70°N)? How can we have a modern bustling city with a well-attended university in Tromsoe, Norway right at 70°N? In same latitude in North America we find ourselves at the Arctic shores of Alaska and Canada, and in Greenland we are on top of an immense ice sheet. In Asia, at 70°N, we are in the northernmost tundra of Siberia, in regions where the ground is always frozen (a phenomenon known as "permafrost").

The Gulf Stream: Norway's Heating System

What is it then that makes Norway so special? The answer is found in the heat transferred by ocean currents and southwesterly winds to the shores and coastal ranges of Norway. Warm waters, heated by the Sun in the subtropics, feed the majestic Gulf Stream, which meanders north off the East Coast to turn east off Newfoundland. The North Atlantic Current then warms the shores of the British Isles and all of northwestern Europe. A portion of this current finds its way up to Iceland and on into the Norwegian Sea. A (relatively) warm current flows all the way up to Tromsoe and beyond, bringing heat into the Barents Sea and even to the shores of Spitsbergen. Thus, Norway has a special heating system, with its origin all the way in the Caribbean!

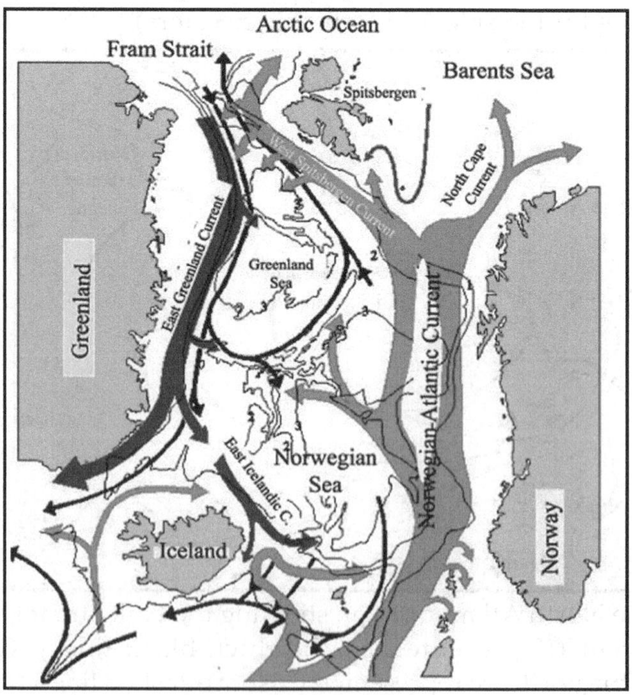

Map of the Norwegian Sea with its complicated currents. Coloured arrows are surface currents and black arrows are deep water currents. Red arrows are "warm" currents originating at the lower latitudes, and blue arrows are "cold" currents originating at higher latitudes. Notice how warm waters tens to found close to Norway, while colder waters are found off of the coast of Greenland.

PHYSICAL EVIDENCE FOR CLIMATIC CHANGE

Evidence for climatic change is taken from a variety of sources that can be used to reconstruct past climates. Reasonably complete global records of surface temperature are available beginning from the mid-late 19th century. For earlier periods, most of the evidence is indirect—climatic changes are inferred from changes in proxies, indicators that reflect climate, such as vegetation, ice cores, dendrochronology, sea level change, and glacial geology.

HISTORICAL AND ARCHAEOLOGICAL EVIDENCE

Climate change in the recent past may be detected by corresponding changes in settlement and agricultural patterns. Archaeological evidence, oral history and historical documents can offer insights into past changes in the climate. Climate change effects have been linked to the collapse of various civilizations.

GLACIERS

Glaciers are considered among the most sensitive indicators of climate change, advancing when climate cools and retreating when climate warms. Glaciers grow and shrink, both contributing to natural variability and amplifying externally forced changes. A world glacier inventory has been compiled since the 1970s, initially based mainly on aerial photographs and maps but now relying more on satellites. This compilation tracks more than 100,000 glaciers covering a total area of approximately 240,000 km^2, and preliminary estimates indicate that the remaining ice cover is around 445,000 km^2. The World Glacier Monitoring Service collects data annually on glacier retreat and glacier mass balance From this data, glaciers worldwide have been found to be shrinking significantly, with strong glacier retreats in the 1940s, stable or growing conditions during the 1920s and 1970s, and again retreating from the mid 1980s to present.

The most significant climate processes since the middle to late Pliocene (approximately 3 million years ago) are the glacial and interglacial cycles. The present interglacial period (the Holocene) has lasted about 11,700 years. Shaped by orbital variations, responses such as the rise and fall of continental ice sheets and significant sea-level changes helped create the climate. Other changes, including Heinrich events, Dansgaard–Oeschger events and the Younger

Dryas, however, show how glacial variations may also influence climate without the orbital forcing.

Glaciers leave behind moraines that contain a wealth of material—including organic matter, quartz, and potassium that may be dated—recording the periods in which a glacier advanced and retreated. Similarly, by tephrochronological techniques, the lack of glacier cover can be identified by the presence of soil or volcanic tephra horizons whose date of deposit may also be ascertained.

VEGETATION

A change in the type, distribution and coverage of vegetation may occur given a change in the climate; this much is obvious. In any given scenario, a mild change in climate may result in increased precipitation and warmth, resulting in improved plant growth and the subsequent sequestration of airborne CO_2. Larger, faster or more radical changes, however, may well result in vegetation stress, rapid plant loss and desertification in certain circumstances.

ICE CORES

Analysis of ice in a core drilled from a ice sheet such as the Antarctic ice sheet, can be used to show a link between temperature and global sea level variations.

The air trapped in bubbles in the ice can also reveal the CO_2 variations of the atmosphere from the distant past, well before modern environmental influences. The study of these ice cores has been a significant indicator of the changes in CO_2 over many millennia, and continues to provide valuable information about the differences between ancient and modern atmospheric conditions.

DENDROCLIMATOLOGY

Dendroclimatology is the analysis of tree ring growth patterns to determine past climate variations. Wide and thick rings indicate a fertile, well-watered growing period, whilst thin, narrow rings indicate a time of lower rainfall and less-than-ideal growing conditions.

POLLEN ANALYSIS

Palynology is the study of contemporary and fossil palynomorphs, including pollen. Palynology is used to infer the geographical distribution of plant species, which vary under different climate conditions. Different groups of plants have pollen with distinctive shapes and surface textures, and since the outer surface of pollen is composed of a very resilient material, they resist decay. Changes in the type of pollen found in different layers of sediment in lakes, bogs, or river deltas indicate changes in plant communities. These changes are often a sign of a changing climate. As an example, palynological studies have been used to track changing vegetation patterns throughout the Quaternary glaciations and especially since the last glacial maximum.

INSECTS

Remains of beetles are common in freshwater and land sediments. Different species of beetles tend to be found under different climatic conditions. Given the extensive lineage of beetles whose genetic makeup has not altered significantly over the millennia, knowledge of the present climatic range of the different species, and the age of the sediments in which remains are found, past climatic conditions may be inferred.

SEA LEVEL CHANGE

Global sea level change for much of the last century has generally been estimated using tide gauge measurements collated over long periods of time to give a long-term average. More recently, altimetre measurements — in combination with accurately determined satellite orbits — have provided an improved measurement of global sea level change. To measure sea levels prior to instrumental measurements, scientists have dated coral reefs that grow near the surface of the ocean, coastal sediments, marine terraces, ooids in limestones, and nearshore archaeological remains. The predominant dating methods used are uranium series and radiocarbon, with cosmogenic radionuclides being sometimes used to date terraces that have experienced relative sea level fall.

CLIMATE CHANGE AND AGRICULTURE

Climate change and agriculture are interrelated processes, both of which take place on a global scale. Global warming is projected to have significant impacts on conditions affecting agriculture, including temperature, carbon dioxide, glacial run-off, precipitation and the interaction of these elements. These conditions determine the carrying capacity of the biosphere to produce enough food for the human population and domesticated animals. The overall effect of climate change on agriculture will depend on the balance of these effects. Assessment of the effects of global climate changes on agriculture might help to properly anticipate and adapt farming to maximize agricultural production.

At the same time, agriculture has been shown to produce significant effects on climate change, primarily through the production and release of greenhouse gases such as carbon dioxide, methane, and nitrous oxide, but also by altering the Earth's land cover, which can change its ability to absorb or reflect heat and light, thus contributing to radiative forcing. Land use change such as deforestation and desertification, together with use of fossil fuels, are the major anthropogenic sources of carbon dioxide; agriculture itself is the major contributor to increasing methane and nitrous oxide concentrations in earth's atmosphere.

IMPACT OF CLIMATE CHANGE ON AGRICULTURE

Despite technological advances, such as improved varieties, genetically modified organisms, and irrigation systems, weather is still a key factor in agricultural productivity, as well as soil properties and natural communities. The effect of climate on agriculture is related to variabilities in local climates rather than in global climate patterns. The Earth's average surface temperature has increased by 1 degree F in just over the last century. Consequently, agronomists consider any assessment has to be individually consider each local area. On the other hand, agricultural trade has grown in recent years, and now provides significant amounts of food, on a national level to major importing countries, as well as comfortable income to exporting ones. The international aspect of trade and security in terms of food implies the need to also consider the effects of climate change on a global scale.

A study published in Science suggests that, due to climate change, "southern Africa could lose more than 30% of its main crop, maize, by 2030. In South Asia losses of many regional staples, such as rice, millet and maize could top 10%". The 2001 IPCC Third Assessment Report concluded that the poorest countries would be hardest hit, with reductions in crop yields in most tropical and sub-tropical regions due to decreased water availability, and new or changed insect pest incidence. In Africa and Latin America many rainfed crops are near their maximum temperature tolerance, so that yields are likely to fall sharply for even small climate changes; falls in agricultural productivity of up to 30% over the 21st century are projected. Marine life and the fishing industry will also be severely affected in some places.

Climate change induced by increasing greenhouse gases is likely to affect crops differently from region to region. For example, average crop yield is expected to drop down to 50% in Pakistan just as to the UKMO scenario whereas corn production in Europe is expected to grow up to 25% in optimum hydrologic conditions.

More favourable effects on yield tend to depend to a large extent on realization of the potentially beneficial effects of carbon dioxide on crop growth and increase of efficiency in water use. Decrease in potential yields is likely to be caused by shortening of the growing period, decrease in water availability and poor vernalization.

In the long run, the climatic change could affect agriculture in several ways:

- Productivity, in terms of quantity and quality of crops
- Agricultural practices, through changes of water use and agricultural inputs such as herbicides, insecticides and fertilizers
- Environmental effects, in particular in relation of frequency and intensity of soil drainage, soil erosion, reduction of crop diversity

- Rural space, through the loss and gain of cultivated lands, land speculation, land renunciation, and hydraulic amenities.
- Adaptation, organisms may become more or less competitive, as well as humans may develop urgency to develop more competitive organisms, such as flood resistant or salt resistant varieties of rice.

They are large uncertainties to uncover, particularly because there is lack of information on many specific local regions, and include the uncertainties on magnitude of climate change, the effects of technological changes on productivity, global food demands, and the numerous possibilities of adaptation.

Most agronomists believe that agricultural production will be mostly affected by the severity and pace of climate change, not so much by gradual trends in climate. If change is gradual, there may be enough time for biota adjustment. Rapid climate change, however, could harm agriculture in many countries, especially those that are already suffering from rather poor soil and climate conditions, because there is less time for optimum natural selection and adaption.

CLIMATE CHANGE AND ECOSYSTEMS

Unchecked global warming could affect most terrestrial ecoregions. Increasing global temperature means that ecosystems will change; some species are being forced out of their habitats because of changing conditions, while others are flourishing. Secondary effects of global warming, such as lessened snow cover, rising sea levels, and weather changes, may influence not only human activities but also the ecosystem.

For the IPCC Fourth Assessment Report, experts assessed the literature on the impacts of climate change on ecosystems.

Rosenzweig *et al.* concluded that over the last three decades, human-induced warming had likely had a discernable influence on many physical and biological systems. Schneider *et al.* concluded, with very high confidence, that regional temperature trends had already affected species and ecosystems around the world. With high confidence, they concluded that climate change would result in the extinction of many species and a reduction in the diversity of ecosystems.

- *Terrestrial ecosystems and biodiversity*: With a warming of 3°C, relative to 1990 levels, it is likely that global terrestrial vegetation would become a net source of carbon. With high confidence, Schneider *et al.* concluded that a global mean temperature increase of around 4°C by 2100 would lead to major extinctions around the globe.
- *Marine ecosystems and biodiversity*: With very high confidence, Schneider *et al.* concluded that a warming of 2°C above 1990 levels would result in mass mortality of coral reefs globally.

- *Freshwater ecosystems*: Above about a 4°C increase in global mean temperature by 2100, Schneider *et al.* concluded, with high confidence, that many freshwater species would become extinct.

Studying the association between Earth climate and extinctions over the past 520 million years, scientists from the University of York write, "The global temperatures predicted for the coming centuries may trigger a new 'mass extinction event', where over 50 per cent of animal and plant species would be wiped out." Many of the species at risk are Arctic and Antarctic fauna such as polar bears and Emperor Penguins. In the Arctic, the waters of Hudson Bay are ice-free for three weeks longer than they were thirty years ago, affecting polar bears, which prefer to hunt on sea ice.

Species that rely on cold weather conditions such as gyrfalcons, and Snowy Owls that prey on lemmings that use the cold winter to their advantage may be hit hard. Marine invertebrates enjoy peak growth at the temperatures they have adapted to, regardless of how cold these may be, and cold-blooded animals found at greater latitudes and altitudes generally grow faster to compensate for the short growing season.

Warmer-than-ideal conditions result in higher metabolism and consequent reductions in body size despite increased foraging, which in turn elevates the risk of predation. Indeed, even a slight increase in temperature during development impairs growth efficiency and survival rate in rainbow trout. Rising temperatures are beginning to have a noticeable impact on birds, and butterflies have shifted their ranges northward by 200 km in Europe and North America. Plants lag behind, and larger animals' migration is slowed down by cities and roads. In Britain, spring butterflies are appearing an average of 6 days earlier than two decades ago.

A 2002 article in Nature surveyed the scientific literature to find recent changes in range or seasonal behaviour by plant and animal species. Of species showing recent change, 4 out of 5 shifted their ranges towards the poles or higher altitudes, creating "refugee species". Frogs were breeding, flowers blossoming and birds migrating an average 2.3 days earlier each decade; butterflies, birds and plants moving towards the poles by 6.1 km per decade. A 2005 study concludes human activity is the cause of the temperature rise and resultant changing species behaviour, and links these effects with the predictions of climate models to provide validation for them. Scientists have observed that Antarctic hair grass is colonizing areas of Antarctica where previously their survival range was limited.

Mechanistic studies have documented extinctions due to recent climate change: McLaughlin *et al.* documented two populations of Bay checkerspot butterfly being threatened by precipitation change. Parmesan states, "Few studies have been conducted at a scale that encompasses an entire species" and McLaughlin *et al.* agreed "few mechanistic studies have linked extinctions to recent climate change." Daniel Botkin and other authors in one study believe that projected rates of extinction are overestimated. Many species of freshwater and saltwater plants and animals are dependent on glacier-fed waters to ensure a cold water habitat that they have adapted to. Some species of freshwater fish need cold water to survive and to reproduce, and this is especially true with Salmon and Cutthroat trout. Reduced glacier run-off can lead to insufficient stream flow to allow these species to thrive. Ocean krill, a cornerstone species, prefer cold water and are the primary food source for aquatic mammals such as the Blue Whale. Alterations to the ocean currents, due to increased freshwater inputs from glacier melt, and the potential alterations to thermohaline circulation of the worlds oceans, may affect existing fisheries upon which humans depend as well.

The white lemuroid possum, only found in the mountain forests of northern Queensland, has been named as the first mammal species to be driven extinct by global warming. The White Possum has not been seen in over three years. These possums cannot survive extended temperatures over 30 °C which occurred in 2005. A final expedition to uncover any surviving White Possums is scheduled for 2009.

FORESTS

Pine forests in British Columbia have been devastated by a pine beetle infestation, which has expanded unhindered since 1998 at least in part due to the lack of severe winters since that time; a few days of extreme cold kill most mountain pine beetles and have kept outbreaks in the past naturally contained. The infestation, which has killed about half of the province's lodgepole pines is an order of magnitude larger than any previously recorded outbreak and passed via unusually strong winds in 2007 over the continental divide to Alberta. An epidemic also started, be it at a lower rate, in 1999 in Colourado, Wyoming, and Montana. The United States forest service predicts that between 2011 and 2013 virtually all 5 million acres of Colourado's lodgepole pine trees over five inches in diametre will be lost.

As the northern forests are a carbon sink, while dead forests are a major carbon source, the loss of such large areas of forest has a positive feedback on global warming. In the worst years, the carbon emission due to beetle infestation of forests in British Columbia alone approaches that of an average year of forest fires in all of Canada or five years worth of emissions from that country's transportation sources.

Besides the immediate ecological and economic impact, the huge dead forests provide a fire risk. Even many healthy forests appear to face an increased risk of forest fires because of warming climates. The 10-year average of boreal forest burned in North America, after several decades of around 10,000 km^2 has increased steadily since 1970 to more than 28,000 km^2 annually. Though this change may be due in part to changes in forest management practices, in the western U.S., since 1986, longer, warmer summers have resulted in a fourfold increase of major wildfires and a sixfold increase in the area of forest burned, compared to the period from 1970 to 1986. A similar increase in wildfire activity has been reported in Canada from 1920 to 1999.

Forest fires in Indonesia have dramatically increased since 1997 as well. These fires are often actively started to clear forest for agriculture. They can set fire to the large peat bogs in the region and the COreleased by these peat bog fires has been estimated, in an average year, to be 15% of the quantity of COproduced by fossil fuel combustion.

MOUNTAINS

Mountains cover approximately 25 per cent of earth's surface and provide a home to more than one-tenth of global human population. Changes in global climate pose a number of potential risks to mountain habitats. Researchers expect that over time, climate change will affect mountain and lowland ecosystems, the frequency and intensity of forest fires, the diversity of wildlife, and the distribution of water.

Studies suggest that a warmer climate in the United States would cause lower-elevation habitats to expand into the higher alpine zone. Such a shift would encroach on the rare alpine meadows and other high-altitude habitats. High-elevation plants and animals have limited space available for new habitat as they move higher on the mountains in order to adapt to long-term changes in regional climate.

Changes in climate will also affect the depth of the mountains snowpacks and glaciers. Any changes in their seasonal melting can have powerful impacts on areas that rely on freshwater run-off from mountains. Rising temperature may cause snow to melt earlier and faster in the spring and shift the timing and distribution of run-off. These changes could affect the availability of freshwater for natural systems and human uses.

ECOLOGICAL PRODUCTIVITY

- A document by Smith and Hitz, it is reasonable to assume that the relationship between increased global mean temperature and ecosystem productivity is parabolic. Higher carbon dioxide concentrations will favourably affect plant growth and demand for water. Higher temperatures could initially be favourable for plant growth. Eventually, increased growth would peak then decline.
- The IPCC, a global average temperature increase exceeding 1.5–2.5°C would likely have a predominantly negative impact on ecosystem goods and services, *e.g.*, water and food supply.
- Research done by the Swiss Canopy Crane Project suggests that slow-growing trees only are stimulated in growth for a short period under higher CO_2 levels, while faster growing plants like liana benefit in the long term. In general, but especially in rainforests, this means

that liana become the prevalent species; and because they decompose much faster than trees their carbon content is more quickly returned to the atmosphere. Slow growing trees incorporate atmospheric carbon for decades.

ARTIFICIAL SEQUESTRATION

For carbon to be sequestered artificially it must first be captured, or it must be significantly delayed or prevented from being re-released into the atmosphere from an existing carbon-rich material, by being incorporated into an enduring usage. Thereafter it can be passively stored or remain productively utilized over time in a variety of ways. For example, upon harvesting, wood can be immediately burned or otherwise serve as a fuel, returning its carbon to the atmosphere, or it can be incorporated into construction or a range of other durable products, thus sequestering its carbon over years or even centuries. One ton of dry wood is equivalent to 1.8 tons of carbon dioxide.

Indeed, a very carefully designed and durable, energy-efficient and energy-capturing building has the potential to sequester as much as or more carbon than was released by the acquisition and incorporation of all its materials and than will be released by building-function "energy-imports" during the structure's existence. Such a structure might be termed "carbon neutral" or even "carbon negative". Building construction and operation are estimated to contribute nearly half of the annual human-caused carbon additions to the atmosphere. Natural-gas purification plants often already have to remove carbon dioxide, either to avoid dry ice clogging gas tankers or to prevent carbon-dioxide concentrations exceeding the 3% maximum permitted on the natural-gas distribution grid. Beyond this, one of the most likely early applications of carbon capture is the capture of carbon dioxide from flue gases at power stations.

A typical new 1000 MW coal-fired power station produces around 6 million tons of carbon dioxide annually. Adding carbon capture to existing plants can add significantly to the costs of energy production; scrubbing costs aside, a 1000 MW coal plant will require the storage of about 50 million barrels of carbon dioxide a year. However, scrubbing is relatively affordable when added to new plants based on coal gasification technology, where it is estimated to raise energy costs for households in the United States using only coal-fired electricity sources from 10 cents per kW·h to 12 cents.

CARBON CAPTURE

Currently, capture of carbon dioxide is performed on a large scale by absorption of carbon dioxide onto various amine-based solvents. Other techniques are currently being investigated, such as pressure swing adsorption, temperature swing adsorption, gas separation membranes, and cryogenics.

Recent pilot studies include flue capture and conversion to baking soda and use of algae for conversion to fuel or feed. In coal-fired power stations, the main alternatives to retrofitting amine-based absorbers to existing power stations are two new technologies: coal gasification combined-cycle and oxy-fuel combustion. Gasification first produces a "syngas" primarily of hydrogen and carbon monoxide, which is burned, with carbon dioxide filtered from the flue gas.

Oxy-fuel combustion burns the coal in oxygen instead of air, producing only carbon dioxide and water vapour, which are relatively easily separated. Some of the combustion products must be returned to the combustion chamber, either before or after separation, otherwise the temperatures would be too high for the turbine. Another long-term option is carbon capture directly from the air using hydroxides. The air would literally be scrubbed of its CO_2 content. This idea offers an alternative to non-carbon-based fuels for the transportation sector. Examples of carbon sequestration at coal plants include converting carbon from smokestacks into baking soda, and algae-based carbon capture, circumventing storage by converting algae into fuel or feed.

OCEANS

Another proposed form of carbon sequestration in the ocean is direct injection. In this method, carbon dioxide is pumped directly into the water at depth, and expected to form "lakes" of liquid CO_2 at the bottom. Experiments carried out in moderate to deep waters indicate that the liquid CO_2 reacts to form solid CO_2 clathrate hydrates, which gradually dissolve in the surrounding waters. This method, too, has potentially dangerous environmental consequences. The carbon dioxide does react with the water to form carbonic acid, H_2CO_3; however, most remains as dissolved molecular CO_2. The equilibrium would no doubt be quite different under the high pressure conditions in the deep ocean. In addition, if deep-sea bacterial methanogens that reduce carbon dioxide were to encounter the carbon dioxide sinks, levels of methane gas may increase, leading to the generation of an even worse greenhouse gas.

The resulting environmental effects on benthic life forms of the bathypelagic, abyssopelagic and hadopelagic zones are unknown. Even though life appears to be rather sparse in the deep ocean basins, energy and chemical effects in these deep basins could have far-reaching implications. Much more work is needed here to define the extent of the potential problems. Carbon storage in or under oceans may not be compatible with the Convention on the Prevention of Marine Pollution by Dumping of Wastes and Other Matter. An additional method of long-term ocean-based sequestration is to gather crop residue such as corn stalks or excess hay into large weighted bales of biomass and deposit it in the alluvial fan areas of the deep ocean basin.

Dropping these residues in alluvial fans would cause the residues to be quickly buried in silt on the sea floor, sequestering the biomass for very long time spans. Alluvial fans exist in all of the world's oceans and seas where river deltas fall off the edge of the continental shelf such as the Mississippi alluvial fan in the gulf of Mexico and the Nile alluvial fan in the Mediterranean Sea. A downside, however, would be an increase in aerobic bacteria growth due to the introduction of biomass, leading to more competition for oxygen resources in the deep sea, similar to the oxygen minimum zone.

GEOLOGICAL SEQUESTRATION

The method of geo-sequestration or geological storage involves injecting carbon dioxide directly into underground geological formations. Declining oil fields, saline aquifers, and unminable coal seams have been suggested as storage sites. Caverns and old mines that are commonly used to store natural gas are not considered, because of a lack of storage safety. CO_2 has been injected into declining oil fields for more than 40 years, to increase oil recovery. This option is attractive because the storage costs are offset by the sale of additional oil that is recovered. Typically, 10-15% additional recovery of the original oil in place is possible.

Further benefits are the existing infrastructure and the geophysical and geological information about the oil field that is available from the oil exploration. Another benefit of injecting CO_2 into Oil fields is that CO_2 is soluble in oil. Dissolving CO_2 in oil lowers the viscosity of the oil and reduces its interfacial tension which increases the oils mobility. All oil fields have a geological barrier preventing upward migration of oil. As most oil and gas has been in place for millions to tens of millions of years, depleted oil and gas reservoirs can contain carbon dioxide for millennia. Identified possible problems are the many 'leak' opportunities provided by old oil wells, the need for high injection pressures and acidification which can damage the geological barrier. Other disadvantages of old oil fields are their limited geographic distribution and depths, which require high injection pressures for sequestration.

A depth of about 1000 m, carbon dioxide is injected as a supercritical fluid, a material with the density of a liquid, but the viscosity and diffusivity of a gas. Unminable coal seams can be used to store CO_2, because CO_2 absorbs to the coal surface, ensuring safe long-term storage. In the process it releases methane that was previously adsorbed to the coal surface and that may be recovered. Again the sale of the methane can be used to offset the cost of the CO_2 storage.

Release or burning of methane would of course at least partially offset the obtained sequestration result–except when the gas is allowed to escape into the atmosphere in significant quantities: methane has a higher global warming potential than CO_2. Saline aquifers contain highly mineralized brines and have so far been considered of no benefit to humans except in a few cases where they have been used for the storage of chemical waste.

Their advantages include a large potential storage volume and relatively common occurrence reducing the distance over which CO_2 has to be transported. The major disadvantage of saline aquifers is that relatively little is known about them compared to oil fields. Another disadvantage of saline aquifers is that as the salinity of the water increases, less CO_2 can be dissolved into aqueous solution. To keep the cost of storage acceptable the geophysical exploration may be limited, resulting in larger uncertainty about the structure of a given aquifer. Unlike storage in oil fields or coal beds, no side product will offset the storage cost. Leakage of CO_2 back into the atmosphere may be a problem in saline-aquifer storage. However, current research shows that several trapping mechanisms immobilize the CO_2 underground, reducing the risk of leakage.

A major research project examining the geological sequestration of carbon dioxide is currently being performed at an oil field at Weyburn in south-eastern Saskatchewan. In the North Sea, Norway's Statoil natural-gas platform Sleipner strips carbon dioxide out of the natural gas with amine solvents and disposes of this carbon dioxide by geological sequestration. Sleipner reduces emissions of carbon dioxide by approximately one million tonnes a year. The cost of geological sequestration is minor relative to the overall running costs. As of April 2005, BP is considering a trial of large-scale sequestration of carbon dioxide stripped from power plant emissions in the Miller oilfield as its reserves are depleted.

In October 2007, the Bureau of Economic Geology at The University of Texas at Austin received a 10-year, $38 million subcontract to conduct the first intensively monitored, long-term project in the United States studying the feasibility of injecting a large volume of CO_2 for underground storage. The project is a research programme of the Southeast Regional Carbon Sequestration Partnership, funded by the National Energy Technology Laboratory of the U.S. Department of Energy. The SECARB partnership will demonstrate CO_2 injection rate and storage capacity in the Tuscaloosa-Woodbine geologic system that stretches from Texas to Florida. Beginning in fall 2007, the project will inject CO_2 at the rate of one million tons per year, for up to 1.5 years, into brine up to 10,000 feet below the land surface near the Cranfield oil field about 15 miles east of Natchez, Mississippi. Experimental equipment will measure the ability of the subsurface to accept and retain CO_2.

MINERAL SEQUESTRATION

Mineral sequestration aims to trap carbon in the form of solid carbonate salts. This process occurs slowly in nature and is responsible for the deposition and accumulation of limestone over geologic time. Carbonic acid in groundwater slowly reacts with complex silicates to dissolve calcium, magnesium, alkalis and silica and leave a residue of clay minerals. The dissolved calcium and magnesium react with bicarbonate to precipitate calcium and magnesium carbonates, a process that organisms use to make shells. When the organisms die, their shells are deposited as sediment and eventually turn into limestone. Limestones have accumulated over billions of years of geologic time and contain much of Earth's carbon. Ongoing research aims to speed up similar reactions involving alkali carbonates. One proposed reaction is that of the olivine-rich rock dunite, or its hydrated equivalent serpentinite with carbon dioxide to form the carbonate mineral magnesite, plus silica and iron oxide. Serpentinite sequestration is favoured because of the non-toxic and stable nature of magnesium carbonate. The ideal reactions involve the magnesium endmember components of the olivine or serpentine, the latter derived from earlier olivine by hydration and silicification. The presence of iron in the olivine or serpentine reduces the efficiency of sequestration, since the iron components of these minerals break down to iron oxide and silica.

STRATOSPHERIC OZONE AND CLIMATEFEEDBACK PROCESSES

The distribution of ozone depends on a balance between chemical processes, which can be affected by changes in the concentration of the ODSs, and transport processes, which can be affected by climate change. Climate change, in its broadest sense, can also affect ozone chemistry directly by modifying the rates of temperature-dependent reactions. In these parts we also discuss the possible impact on ozone of changing stratospheric water vapour abundances, which may be regulated by climate. Changes in stratospheric ozone can also affect climate. As well as the direct impact of ozone as an absorber of UV radiation and as a greenhouse gas, there are other effects that could be important. For example, changes in ozone could affect the lifetime of reactive greenhouse gases, by changing the penetration of UV radiation. Changes in the structure of the stratosphere, caused by ozone changes, could alter the interaction between the troposphere and stratosphere and lead to further changes in stratospheric ozone. Improved knowledge of these various feedback processes is essential for informing the numerical models used to predict future chemistry-climate interactions.

Note that other factors, which are not discussed in detail here, can also affect the interaction between ozone and climate. Perhaps the most obvious is the impact of major volcanic eruptions. These can lead to an increase in volcanic aerosol in the stratosphere, which can influence both climate and the chemical processes controlling the ozone layer.

IMPACT OF ODSS ON STRATOSPHERIC OZONE

The abundance of ozone in the stratosphere at a particular location is governed by three processes: photochemical production, destruction by catalytic cycles, and transport processes. Photochemical production in the stratosphere occurs mostly through the photolysis of O_2, with loss via catalytic cycles involving hydrogen, nitrogen, and halogen radicals. The relative importance of the various loss cycles in the stratosphere varies substantially with altitude. Above about 45 km, loss through HOx dominates, while below this altitude NOx-catalyzed ozone loss is most important. The importance of the ClOx-catalyzed ozone loss cycle, which varies with chlorine loading, peaks at about 40 km. Below about 25 km HOx-driven ozone loss cycles dominate again.

Upper Stratosphere

In the upper stratosphere the ozone budget is largely understood, although uncertainties remain regarding the rate constants of key radical reactions. In particular, strong evidence has been accumulated that the observed ozone depletion in the upper stratosphere is caused by increased levels of stratospheric chlorine as originally proposed by Molina and Rowland and Crutzen.

Because of the direct correspondence between the stratospheric chlorine abundance and ozone depletion in the upper stratosphere, it has been suggested that the response of stratospheric ozone to the declining stratospheric chlorine levels might be first detectable in the upper stratosphere.

Polar Regions

In recent decades stratospheric ozone losses have been most pronounced in polar regions during winter and spring.

These losses are determined largely by three chemical factors:

1. The conversion of chlorine reservoirs into active, ozone-destroying forms through heterogeneous reactions on the surfaces of polar stratospheric cloud particles;
2. The availability of sunlight that drives the catalytic photochemical cycles that destroy ozone;
3. The timing of the deactivation of chlorine.

emperature controls the formation and destruction of polar stratospheric clouds and thus the timing of activation and deactivation of chlorine. Furthermore, temperature controls the efficiency of the catalytic cycles that destroy ozone in the presence of sunlight. However, polar ozone loss would not occur without a prominent dynamical feature of the stratosphere in winter and spring: the polar vortex. In both hemispheres, the polar vortex separates polar air from mid-latitude air to a large extent, and within the vortex the low-temperature conditions that develop are the key factor for polar ozone loss. These two factors are dynamically related: a strong vortex is generally also a colder vortex. The two crucial questions for future polar ozone are whether, in an increased greenhouse-gas climate, the region of low stratospheric temperatures will increase in area and whether it will persist for longer in any given year. When anthropogenically emitted ODSs are eventually removed from the stratosphere, the stratospheric halogen burden will be much lower than it is today and will be controlled by the naturally occurring source gases methyl chloride and methyl bromide.

Under such conditions, dramatic losses of polar ozone in winter and spring as we see today are not expected to occur. However, the rate of removal of anthropogenic halogens from the atmosphere depends on the atmospheric lifetimes of CFCs and is considerably slower than the rate at which halogens have been increasing in the decades prior to approximately the year 2000, when the stratospheric halogen loading peaked. A complete removal of anthropogenic halogens from the atmosphere will take more than a century. During this period of enhanced levels of halogens caused by past anthropogenic emissions, the polar stratosphere will remain vulnerable to climate perturbations, such as increasing water vapour or a cooling of the stratosphere, that lead to enhanced ozone destruction.

Lower-Stratospheric Mid-Latitudes

It is well established that ozone in the lower stratosphere at midlatitudes has been decreasing for a few decades; both measurements of ozone locally in the lower stratosphere and measurements of column ozone show a clear decline. However, because mid-latitude ozone loss is much less severe than polar ozone loss, it cannot be identified in measurements in any one year but is rather detected as a downward trend in statistical analyses of longer time-series.

It is clear that chemical loss driven by halogens is very important, but other possible effects that may contribute to the observed mid-latitude trends have also been identified. A definitive quantitative attribution of the trends to particular mechanisms has not yet been achieved. WMO reviewed this issue most recently; as suggested, not repeat that detailed analysis here but instead provide a brief summary. The chemical processes that may affect trends of mid-latitude ozone are essentially related to the ODS trends that are known to be responsible for the observed ozone loss in the upper stratosphere and in the polar regions.

Halogen chemistry may lead to ozone depletion in the mid-latitude lower stratosphere through a number of possible mechanisms, including:

- Export of air that has encountered ozone destruction during the winter from the polar vortex.
- Export of air with enhanced levels of active chlorine from the polar vortex.
- In situ activation of chlorine either on cold liquid sulfate aerosol particles or on ice particles. Further, the reaction of N_2O_5 with water on liquid aerosol particles at higher temperatures indirectly enhances the concentrations of ClO at mid-latitudes.

- Ozone depletion due to elevated levels of BrO in the lower stratosphere, possibly caused by transport of very shortlived halogen-containing compounds or BrO across the tropopause.

The first mechanism results from transport combined with polar ozone loss, whereas the remaining three mechanisms all involve in situ chemical destruction of ozone at mid-latitudes.

All four mechanisms are ultimately driven by the increase of halogens in the atmosphere over the past decades.

Thus, while Millard *et al.* emphasized the strong interannual variability in the relative contributions of the different mechanisms to the seasonal mid-latitude ozone loss during the winter-spring period, they nonetheless showed that ozone loss driven by catalytic cycles involving halogens was always an important contributor to the simulated mid-latitude ozone loss in the five winters in the 1990s that they studied.

Similarly, Chipperfield found that the observed midlatitude column ozone decrease from 1980 to the early 1990s could be reproduced in long-term simulations in a 3-D chemistry- transport model. The modelled ozone is affected by dynamical interannual variability, but the overall decreases are dominated by halogen trends; and about 30-50% of the modelled halogen-induced change is a result of high latitude processing on PSCs. Under climate change the strength of the polar vortex may change but the sign of this change is uncertain.

A change in the strength and temperature of the vortex will affect chlorine activation and ozone loss there and, through mechanisms and ozone loss in mid-latitudes. The possibility that chlorine might be activated on cirrus clouds or on cold liquid aerosol particles in the lowermost stratosphere was revisited recently by Bregman *et al.* Based on their model results it seems unlikely that this process is the main mechanism for the observed long-term decline of ozone in the mid-latitude lower stratosphere. Very short-lived organic chlorine-, bromine-, and iodine-containing compounds possess a potential to deplete stratospheric ozone.

However a quantitative assessment of their impact on stratospheric ozone is made difficult by their short lifetime, so there is need to consider the transport pathways from the troposphere to the stratosphere of these compounds in detail.

An upper limit for total stratospheric iodine, Iy, of 0.10 ± 0.02 ppt was recently reported by Bösch *et al.* The impact of this magnitude of iodine loading on stratospheric ozone is negligible. The disturbance of the mid-latitude ozone budget caused by anthropogenic emissions of ODSs will ultimately cease when the stratospheric halogen burden has reached low enough levels. However, like polar ozone, mid-latitude ozone will for many decades remain vulnerable to an enhancement of halogen-catalyzed ozone loss caused by climate change and by natural phenomena such as volcanic eruptions.

IMPACT OF TEMPERATURE CHANGES ONOZONE CHEMISTRY

The increasing abundance of WMGHGs in the atmosphere is expected to lead to an increase in temperature in the troposphere. Furthermore, increasing concentrations of most of these gases, notably CO_2, N_2O and CH4, are expected to lead to a temperature decrease in the stratosphere. By far the strongest contribution to stratospheric cooling from the WMGHGs comes from CO_2. Temperatures in the stratosphere are also believed to have decreased in part because of the observed reductions in ozone concentrations; any such cooling would have a feedback on the ozone changes. Decreasing stratospheric temperatures lead to a reduction of the ozone loss rate in the upper stratosphere, thereby indirectly leading to more ozone in this region. This reduction of ozone loss is caused by the very strong positive temperature dependence of the ozone loss rate, mainly owing to the Chapman reactions and the NOx cycle. This inverse relationship between ozone and temperature changes in the upper stratosphere has been known for many years. More recently, differences in temperature between the two hemispheres have been identified as a cause of inter-hemispheric differences in both the seasonal cycle of upper-stratospheric ozone abundances and in the upper-stratospheric ozone trend deduced from satellite measurements. In the mid-latitude lower stratosphere, one mechanism that leads to ozone loss and is directly sensitive to temperature changes involves heterogeneous reactions on the surfaces of cloud and cold aerosol particles. The rates of many of these reactions increase strongly with decreasing temperature. Similarly, the reaction rates increase with increasing water-vapour concentrations, so that future increases in water, should they occur, would also have an impact.

Note, however, that in the lower-stratospheric mid-latitudes the most important heterogeneous reactions are hydrolysis of N_2O_5 and bromine nitrate, which are relatively insensitive to temperature and water vapour concentrations. Therefore, the expected impact of climate change on the chemical mechanism is expected to be relatively small. In addition to this impact on heterogeneous chemistry, Zeng and Pyle have argued that a reduction in temperature in the lower stratosphere can slow the rate of HOx-driven ozone destruction. Polar ozone loss occurs when temperatures in a large enough region sink below the threshold temperature for the existence of PSCs because chlorine is activated by heterogeneous reactions on the surfaces of PSCs. Therefore, a cooling of the stratosphere enhances Arctic ozone loss if the volume of polar air with temperatures below the PSC threshold value increases.

Moreover, a stronger PSC activity is expected to lead to a greater denitrification of the Arctic vortex, hence a slower deactivation of chlorine and, consequently, a greater chemical ozone loss. Rex *et al.* have deduced an empirical relation between observed temperatures and observed winter-spring chemical loss of Arctic ozone. Based on this relation, and for current levels of chlorine, about 15 DU additional loss of total ozone is expected for each 1 K cooling of the Arctic lower stratosphere. Although the lower stratosphere is expected to generally cool with increasing greenhouse-gas concentrations, the temperature changes in the lower stratosphere during the polar wintertime will be sensitive to any change in circulation associated with dynamical feedbacks, principally from changes in planetary wave drag. In principle, these feedbacks could be of either sign and thus could lead to enhanced cooling or even to warming.

An early model study found enhanced cooling, to the extent that polar ozone loss would be expected to increase during the next 10 to 15 years even while halogen levels decreased. More recent studies with higher-resolution models have found a much less dramatic dynamical feedback, with some models showing an increase and some a decrease in planetary wave drag, and with all models predicting a relatively small change in Arctic ozone over the next few decades. However, it should be noted that model results for the Arctic are difficult to assess because the processes leading to polar ozone depletion show so much natural variability that the atmosphere may evolve anywhere within the envelope provided by an ensemble of model simulations. In any event, it is the Arctic that is most sensitive to the effects caused by climate change. Present-day temperatures in the Arctic lie close to the threshold value for the onset of heterogeneous chlorine activation and thus close to the threshold value for the onset of rapid ozone loss chemistry; in the Antarctic temperatures are much lower and thus ozone loss is not as sensitive to temperature changes.

IMPACT OF METHANE AND WATER VAPOUR CHANGES ON OZONE CHEMISTRY

With the exception of the high latitudes in winter, ozone in the upper stratosphere is in a photochemical steady-state; photochemical reactions are sufficiently fast that ozone concentrations are determined by a local balance between photochemical production and loss. Nonetheless, transport has a significant indirect influence on upper stratospheric ozone insofar as it determines the concentrations of trace compounds such as CH_4, H_2O, CFCs and N_2O, all of which act as precursors of the radicals that determine the ozone chemistry. Furthermore, CH_4 is of particular importance because it is the primary mechanism for the conversion of reactive Cl to the unreactive HCl reservoir via the reaction $CH_4 + Cl \rightarrow HCl + CH_3$, and thus affects the efficiency of the chlorine-driven ozone loss. Changes in upper stratospheric CH_4 have important implications for upper stratospheric ozone.

Siskind *et al.* found that for the period 1992–1995, the increase in active chlorine resulting from the CH_4 decrease was the largest contributor to the ozone changes occurring over that time. Indeed, a measured increase in upper-stratospheric ClO between 1991 and 1997, which is significantly greater than that expected from the increase in the chlorine source gases alone, may be explained by the observed concurrent decrease of CH_4. Further, Li *et al.* find that inter-hemispheric differences in CH_4 are partly responsible for observed differences in upper stratospheric ozone trends between the two hemispheres. Stratospheric water vapour is the primary source of the HOx radicals that drive the dominant ozone loss cycles in the upper stratosphere. An increase in stratospheric water vapour is therefore expected to lead to a greater chemical loss of ozone in the upper stratosphere. This effect has been investigated quantitatively in model simulations.

In a study of conditions for the year 2010, Jucks and Salawitch find that above 45 km an increase of 1% in the water vapour mixing ratio would completely negate the increase of ozone driven by the 15% decrease of inorganic chlorine that is expected by the year 2010. At 40 km the increase of ozone would still be reduced by about 50%. Shindell conducted a general circulation model study for the period 1979–1996. The simulations show a significant chemical effect of water vapour increases on ozone concentrations, with a reduction of more than 1% between 45 and 55 km, and a maximum impact of about 4% at 50 km. Further, Li *et al.* found that the annually averaged ozone trend at 45°S and 1.8 hPa increases by 1% per decade for a water vapour increase of 1% yr–1. These model results point to an anticorrelation between ozone and water vapour in the upper stratosphere and lower mesosphere. However, such an anticorrelation is not seen in observations and ozone in this region varies much less than predicted by models.

In the mid-latitude lower stratosphere, any increase in water vapour would also be expected to lead to ozone loss because of an intensified HOx ozone loss cycle. The model results of Dvortsov and Solomon predict that an increase in stratospheric water vapour of 1% yr–1 translates to an additional depletion of mid-latitude column ozone by 0.3% per decade.

In the polar lower stratosphere, the rates of the heterogeneous reactions that are responsible for the activation of chlorine increase with increasing water vapour concentrations. An increase in stratospheric water vapour essentially means that the temperature threshold at which PSCs form, and thus heterogeneous reactions rates begin to become significant for chlorine activation, is shifted to higher values.

If stratospheric water vapour were to increase it could lead to a substantially enhanced Arctic ozone loss in the future. In considering concerning stratospheric water vapour, it needs to be borne in mind that there is considerable uncertainty about the sign of future water vapour changes in light of the puzzling past record.

THE ROLE OF TRANSPORT FOR OZONE CHANGES

Transport is a key factor influencing the seasonal and interannual variability of stratospheric ozone. Seasonal variations in transport force the large winter-spring build-up of extra-tropical total ozone in both hemispheres, and inter-hemispheric differences in transport cause corresponding differences in extra-tropical total ozone. In both hemispheres, mid-latitude ozone decreases during summer and returns to approximate photochemical balance by autumn, and there is a strong persistence of the dynamically forced anomalies throughout summer. The interannual variability in the winter-spring build-up is greater in the NH, reflecting the greater dynamical variability of the NH stratosphere.

Stratospheric Planetary-Wave-induced Transport and Mixing

Large-scale transport of ozone is a result of advection by the Brewer-Dobson circulation and of eddy transport effects; both of these mechanisms are first-order terms in the zonal mean ozone transport equation. The strength of the Brewer-Dobson circulation is directly tied to dissipating planetary waves forced from the troposphere, and eddy transports of ozone are also linked to planetary wave activity, so that net ozone transport is tied to the variability of forced planetary waves. The amount of dissipating wave activity within the stratosphere is related to the vertical component of wave activity entering the lower stratosphere.

This quantity can be derived from conventional meteorological analyses and is a convenient proxy used to quantify PWD. A significant correlation has been found between interannual changes in PWD and total ozone build-up during winter and spring. The fact that the effect of PWD on ozone is seasonal and has essentially no interannual memory suggests that long-term changes in PWD can be expected to lead to long-term changes in winter-spring ozone, all else being equal. However, although the basic physics of the ozone-PWD connection is well understood, its quantification via correlations is at best crude, and this limits our ability to attribute changes in ozone to changes in PWD. In the NH, there have been interannual variations in various meteorological parameters during the period 1980–2000 that together paint a fairly consistent, albeit incomplete, picture. During the mid-1990s the NH exhibited a number of years when the Arctic wintertime vortex was colder and stronger and more persistent.

Any dynamically induced component of these changes requires a weakened Brewer-Dobson circulation, which in turn requires a decrease in PWD. Such a decrease in PWD during this period has been documented, although the results were sensitive to which months and time periods were considered. Randel *et al.* show that for the period 1979–2000, PWD in the NH increased during early winter and decreased during mid-winter. This seasonal variation is consistent with the Arctic early winter warming and late winter cooling seen over the same time period at 100 hPa. The weakened Brewer-Dobson circulation during midwinter implies a decrease in the winter build-up of mid-latitude ozone, and Randel *et al.* estimated that the decreased wave driving may account for about 20–30% of the observed changes in ozone in the January to March period. Changes in SH dynamics are not as clear as in the NH, primarily because meteorological reanalysis data sets are less well constrained by observations. Planetary waves, in addition to affecting the temperature and chemistry of the polar stratosphere through the processes described earlier, can also displace the centre of the polar vortex off the pole.

This has important implications for ozone and NOx chemistry because air parcel trajectories within the vortex are then no longer confined to the polar night but experience short periods of sunlight. Polar ozone loss processes are usually limited by the poleward retreat of the terminator. Planetary wave distortion of the vortex can expose deeper vortex air to sunlight and cause ozone depletion chemistry to start earlier than it would have otherwise. In this way, waveinduced displacements of the vortex can drive a mid-winter start to Antarctic ozone depletion. Because wave-induced forcing in the stratosphere is believed to come primarily from planetary-scale Rossby waves that are generated in the troposphere during wintertime, future changes in the generation of tropospheric waves may influence polar ozone abundance. However there is as yet no consensus from models on the sign of this change.

Tropopause Variations and Ozone Mini-Holes

Tropospheric circulation and tropopause height variations also affect the mid-latitude distribution of column ozone. The relationship between local mid-latitude tropopause height and column ozone on day-to-day time scales is well documented. Day-to-day changes in tropopause height are associated with the passage of synoptic-scale disturbances in the upper troposphere and lower stratosphere, which affect ozone in the lower stratosphere through transport and can result in large local changes in column ozone, particularly in the vicinity of storm tracks. In extreme situations, they can lead to socalled ozone mini-holes, which occur over both hemispheres and have the lowest column ozone levels observed outside the polar vortices, sometimes well under 220 DU. Ozone mini-holes do not primarily entail a destruction of ozone, but rather its re-distribution. Changes in the spatial and temporal occurrence of synoptic-scale processes and Rossby wave breaking that are induced by climate change or natural variability, could lead to changes in the distribution or frequency of the occurrence of mini-hole and low-ozone events, and thus to regional changes in the mean column ozone. Observations have shown that over the NH the altitude of the extra-tropical tropopause has generally increased over recent decades. Radiosonde measurements over both Europe and Canada show an increase in altitude of about 300–600 m over the past 30 years, with the precise amount depending on location. Consistent regional increases are also seen in meteorological reanalyses over both hemispheres, although reanalysis trend studies should be viewed with care. Spatial patterns show increases over the NH and SH beginning at mid-latitudes and reaching a maximum towards the poles. The magnitude of the changes can also depend on the longitude.

However, the effects of tropospheric circulation changes on ozone and the relation of both to tropopause height changes remain poorly understood at present. In particular, the relationship between tropopause height and ozone mentioned earlier applies to single stations; there is no reason to expect it to apply in the zonal mean, or on longer time scales over which ozone transport is irreversible. Because of our poor understanding of what controls the zonal-mean midlatitude tropopause height, or whether it is possible to consider such processes in a zonal-mean approach, it is not clear that the ozone-tropopause height correlations can be extended to decadal time scales in order to estimate changes in ozone.

Some recent model studies suggest that lower stratospheric cooling caused by ozone depletion, acting together with tropospheric warming due to WMGHGs, can be a contributor to tropopause height changes in which case the tropopause height changes cannot be entirely considered as a cause of the ozone changes. One source of uncertainty in these model calculations is that most climate models cannot resolve the observed tropopause height changes, so these must be inferred by interpolation; another is that the observed ozone changes are not well quantified close to the tropopause. Furthermore there remain considerable differences between the various reanalysis products available, as well as between the model results.

Stratosphere-Troposphere Exchange

Stratosphere-troposphere exchange processes also affect the ozone distribution and have the capability to affect tropospheric chemistry. STE is a two-way process that encompasses transport from the troposphere into the stratosphere and from the stratosphere into the troposphere through a variety of processes. Some aspects of STE are implicit. The net mass flux from the troposphere into the stratosphere and back again is driven by PWD through the Brewer-Dobson circulation. Although the global approach can explain the net global flux, local and regional processes need to be identified and assessed to fully understand the distribution of trace species being exchanged. Such understanding is relevant both for chemical budgets and for climate change and variability studies. Although there is a net TST in the tropics and a net STT in the extratropics, a number of mechanisms that lead to STT in the tropics and TST in the extratropics have been identified. In the extratropics TST can significantly affect the composition of the lowermost stratosphere, even if the exchange cannot reach higher into the stratosphere, by introducing, during deep TST events, near-surface pollutants, and ozone-poor and humid air.

Similarly, deep STT events can transport ozone-rich air into the mid- and lower troposphere. It should be noted, however, that so-called shallow events, with exchanges near the tropopause, remain the main feature in the extratropics. Regions of occurrence of such exchange processes, at least in the NH, appear to be linked to the storm tracks and their variability, that is, in the same region where mini-holes are most frequent. The past and future variability of STE remains to be assessed. It is clear that changes in planetary wave activity will modify the Brewer-Dobson circulation and hence affect global transport processes. As for regional mechanisms, seasonal and interannual variability, driven, for example, by the North Atlantic Oscillation and the El Niño-Southern Oscillation have already been observed. Changes in the occurrence of regional and local weather phenomena could also modify STE. Fifteen-year studies with reanalysis products have not yielded any distinct trends. Given the discrepancies between different approaches to evaluate STE and inhomogeneities in meteorological analyses and reanalyses, consistent trend studies are not available as of yet.

The Tropical Tropopause Layer

Air enters the stratosphere primarily in the tropics, and hence the physical and chemical characteristics of air near the tropical tropopause behave as boundary conditions for the global stratosphere. The tropical tropopause is relatively high, near 17 km. The tropospheric lapse rate is determined by radiative-convective equilibrium, whereas the thermal structure above 14 km is primarily in radiative balance, which is characteristic of the stratosphere. Overall, the region of the tropical atmosphere between about 12 km and the tropopause has characteristics intermediate to those of the troposphere and stratosphere, and is referred to as the tropical tropopause layer. Thin cirrus clouds are observed over large areas of the TTL, although their formation mechanism(s) and effects on large-scale circulation are poorly known. In the tropics, the background clear-sky radiative balance shifts from cooling in the troposphere to heating in the stratosphere, with the transition occurring at around 15 km. The region of heating above about 15 km is linked to mean upward motion into the lower stratosphere, and this region marks the base of the stratospheric Brewer-Dobson circulation.

The TTL is coupled to stratospheric ozone through its control of stratospheric water vapour and by the transport of tropospheric source gases and ODSs into the lower stratosphere, so changes in the TTL could affect the stratospheric ozone layer. If the residence time of air in the TTL before it is transported into the stratosphere is short, then it may be possible that short-lived halogen compounds or their degradation products could enter the lower stratosphere and contribute to ozone loss there. Air entering the stratosphere is dehydrated as it passes through the cold tropical tropopause, and this drying accounts for the extreme aridity of the global stratosphere. Furthermore, the seasonal cycle in tropopause temperature imparts a strong seasonal variation in stratospheric water vapour, which then propagates with the mean stratospheric transport circulation.

Year-to-year variations in tropical tropopause temperatures are also highly correlated with global stratospheric water vapour anomalies, although there remain some issues concerning the consistency of decadal-scale changes in water vapour. However, while there is strong empirical coupling between tropopause temperatures and stratospheric water vapour, details of the dehydration mechanism(s) within the TTL are still a topic of scientific debate. One critical, unanswered question is whether, and how, the TTL will change in response to climate change and what will be the resulting impact on stratospheric ozone.

OZONE IN THE ATMOSPHERE AND ITS ROLE IN CLIMATE

A number of important concepts concerning the ozone layer and its role in the stratosphere. Ozone, like water vapour and carbon dioxide, is an important and naturally occurring greenhouse gas; that is, it absorbs and emits radiation in the thermal infrared, trapping heat to warm the Earth's surface. In contrast to the so-called wellmixed greenhouse gases, stratospheric ozone has two distinguishing properties. First, its relatively short chemical lifetime means that it is not uniformly mixed throughout the atmosphere and therefore its distribution is controlled by both dynamical and chemical processes. In fact, unlike the WMGHGs, ozone is produced entirely within the atmosphere rather than being emitted into it. Second, it is a very strong absorber of short wavelength UV radiation.

The ozone layer's absorption of this UV radiation leads to the characteristic increase of temperature with altitude in the stratosphere and, in consequence, to a strong resistance to vertical motion. As well as ozone's role in climate it also has more direct links to humans: its absorption of UV radiation protects much of Earth's biota from this potentially damaging short wavelength radiation. In contrast to the benefits of stratospheric ozone, high surface ozone values are detrimental to human health. The distribution of ozone in the atmosphere is maintained by a balance between photochemical production and loss, and by transport between regions of net production and net loss. A number of different chemical regimes can be identified for ozone. In the upper stratosphere, the ozone distribution arises from a balance between production following photolysis of molecular oxygen and destruction via a number of catalytic cycles involving hydrogen, nitrogen and halogen radical species. The halogens arise mainly from anthropogenic ODSs. In the upper stratosphere, the rates of ozone destruction depend on temperature and on the concentrations of the radical species. In the lower stratosphere, reactions on aerosols become important. The distribution of the radicals can be affected by heterogeneous and multiphase chemistry acting on condensed matter. At the low temperature of the wintertime polar lower stratosphere, this is the chemistry that leads to the ozone hole.

The large-scale circulation of the stratosphere, known as the Brewer-Dobson circulation, systematically transports ozone poleward and downward. Because ozone photochemical reactions proceed quickly in the sunlit upper stratosphere, this transport has little effect on the ozone distribution there as ozone removal by transport is quickly replenished by photochemical production. However, this transport leads to significant variations of ozone in the extra-tropical lower stratosphere, where the photochemical relaxation time is very long and ozone can accumulate on seasonal time scales. Due to the seasonality of the Brewer- Dobson circulation, ozone builds up in the extra-tropical lower stratosphere during winter and spring through transport, and then decays photochemically during the summer when transport is weaker. The columnozone distribution is dominated by its distribution in the lower stratosphere and reflects this seasonality. Furthermore, planetary waves are stronger in the Northern Hemisphere than in the Southern Hemisphere, because of the asymmetric distribution of the surface features that, in combination with surface winds, force the waves.

The stratospheric Brewer- Dobson circulation is stronger during the NH winter, and the resulting extra-tropical build-up of ozone during the winter and spring is greater in the NH than in the SH. Variations in the Brewer-Dobson circulation also influence polar temperatures in the lower stratosphere; stronger wave forcing coincides with enhanced circulation and higher polar temperatures. Since temperature affects ozone chemistry, dynamical and chemical effects on column ozone thus tend to act in concert and are coupled. Human activities have led to changes in the atmospheric concentrations of several greenhouse gases, including tropospheric and stratospheric ozone and ODSs and their substitutes.

Changes to the concentrations of these gases alter the radiative balance of the Earth's atmosphere by changing the balance between incoming solar radiation and outgoing infrared radiation. Such an alteration in the Earth's radiative balance is called a radiative forcing. This report, past IPCC reports and climate change protocols have universally adopted the concept of radiative forcing as a tool to gauge and contrast surface climate change caused by different mechanisms. Positive radiative forcings are expected to warm the Earth's surface and negative radiative forcings are expected to cool it. Changes in carbon dioxide provide the largest radiative forcing term and are expected to be the largest overall contributor to climate change. In contrast with the positive radiative forcings due to increases in other greenhouse gases, the radiative forcing due to stratospheric ozone depletion is negative. Halocarbons are particularly effective greenhouse gases in part because they absorb the Earth's outgoing infrared radiation in a spectral range where energy is not removed by carbon dioxide or water vapour.

Halocarbon molecules can be many thousands of times more efficient at absorbing the radiant energy emitted from the Earth than a molecule of carbon dioxide, which explains why relatively small amounts of these gases can contribute significantly to radiative forcing of the climate system.

Because halocarbons have low concentrations and absorb in the atmospheric window, the magnitude of the direct radiative forcing from a halocarbon is given by the product of its tropospheric mixing ratio and its radiative efficiency. In contrast, for the more abundant greenhouse gases there is a nonlinear relationship between the mixing ratio and the radiative forcing. Since 1970 the growth in halocarbon concentrations and the changes in ozone concentrations have been very significant contributors to the total radiative forcing of the Earth's atmosphere.

Because halocarbons have likely caused most of the stratospheric ozone loss, there is the possibility of a partial offset between the positive forcing of halocarbon that are ODSs and the negative forcing from stratospheric ozone loss.

The climate impacts of ozone changes are not confined to the surface: stratospheric ozone changes are probably responsible for a significant fraction of the observed cooling in the lower stratosphere over the last two decades and may alter atmospheric dynamics and chemistry. Further, it was predicted that depletion of stratospheric ozone would lead to a global increase in erythemal UV radiation at the surface of about 3%, with much larger increases at high latitudes; these predicted high latitude increases in surface UV dose have indeed been observed.

CHAPTER OUTLINE

Our aim in this stage is to review the scientific understanding of the interactions between ozone and climate. We interpret the term 'climate' broadly, to include stratospheric temperature and circulation, and refer to tropospheric effects as 'tropospheric climate'.

This is a broad topic; our review will take a more restricted view concentrating on the interactions as they relate to stratospheric ozone and the role of the ODSs and their substitutes. For this reason the role of tropospheric ozone in the climate system is mentioned only briefly. Similarly, some broader issues, including possible changes in stratospheric water vapour and climate-dependent changes in biogenic emissions will not be discussed in any detail. The relationship between ozone and the solar cycle is well established, and the solar cycle has been used as an explanatory variable in ozone trend analysis. An update on stratospheric observations of ozone, ozone-related species and temperature. A process level, the various feedbacks connecting stratospheric ozone and the climate system. The understanding of these processes leads to a discussion of the attribution of past changes in ozone, the prediction of future changes, and the connection of both with climate. Finally, reviews the trade-offs between ozone depletion and radiative forcing, focusing on ODSs and their substitutes.

OBSERVED CHANGES IN THE STRATOSPHERE

OBSERVED CHANGES IN STRATOSPHERIC OZONE

Global and hemispheric-scale variations in stratospheric ozone can be quantified from extensive observational records covering the past 20 to 30 years. There are numerous ways to measure ozone in the atmosphere, but they fall broadly into two categories: measurements of column ozone and measurements of the vertical profile of ozone. Approximately 90% of the vertically integrated ozone column resides in the stratosphere. There are more independent measurements, longer time-series, and better global coverage for column ozone. Regular measurements of column ozone are available from a network of surface stations, mostly in the mid-latitude NH, with reasonable coverage extending back to the 1960s. Near-global, continuous column ozone data are available from satellite measurements beginning in 1979. The different observational data sets can be used to estimate past ozone changes, and the differences between data sets provide a lower bound of overall uncertainty.

The differences indicate good overall agreement between different data sources for changes in column ozone, and thus we have reasonable confidence in describing the spatial and temporal characteristics of past changes. Five data sets of zonal and monthly mean column ozone values developed by different scientific teams were used to quantify past ozone changes; they include ground-based measurements covering 1964–2001, and several different satellite data sets extending in time over 1979–2001. The analyses first remove the seasonal cycle from each data set and the deviations are area weighted and expressed as anomalies with respect to the period 1964–1980. The global ozone amount shows decreasing values between the late 1970s and the early 1990s, a relative minimum during 1992–1994, and slightly increasing values during the late 1990s.

Global ozone for the period 1997–2004 was approximately 3% below the 1964–1980 average values. Since systematic global observations began in the mid-1960s, the lowest annually averaged global ozone occurred during 1992–1993. These changes are evident in each of the available global data sets. No significant long-term changes in column ozone have been observed in the tropics. Column ozone changes averaged over mid-latitudes are significantly larger in the SH than in the NH; averaged for the period 1997–2001, SH values are 6% below pre-1980 values, whereas NH values are 3% lower. Also, there is significant seasonality to the NH mid-latitude losses whereas long-term SH losses are about 6% year round. The most dramatic changes in ozone have occurred during the spring season over Antarctica, with the development during the 1980s of a phenomenon known as the ozone hole.

The ozone hole now recurs every spring, with some interannual variability and occasional extreme behaviour. In most years the ozone concentration is reduced to nearly zero over a layer several kilometres deep within the lower stratosphere in the Antarctic polar vortex. Since the early 1990s, the average October column ozone poleward of 63°S has been more than 100 DU below pre-ozone-hole values with up to a 70% local decrease for periods of a week or so. Compared with the Antarctic, Arctic ozone abundance in the winter and spring is highly variable because of interannual variability in chemical loss and in dynamical transport. Dynamical variability within the winter stratosphere leads to changes in ozone transport to high latitudes, and these transport changes are correlated with polar temperature variability–with less ozone transport being associated with lower temperatures. Low temperatures favour halogen-induced chemical ozone loss.

Thus, in recent decades, halogen-induced polar ozone chemistry has acted in concert with the dynamically induced ozone variability, and has led to Arctic column ozone losses of up to 30% in particularly cold winters. In particularly dynamically active, warm winters, however, the estimated chemical ozone loss has been very small. Changes in the vertical profile of ozone are derived primarily from satellites, ground-based measurements and balloon-borne ozonesondes. Long records from ground-based and balloon data are available mainly for stations over NH mid-latitudes. Ozone profile changes over NH mid-latitudes exhibit two maxima, with decreasing trends in the upper stratosphere and in the lower stratosphere during the period 1979–2000. Ozone profile trends show a minimum near 30 km. The vertically integrated profile trends are in agreement with the measured changes in column ozone.

OBSERVED CHANGES IN ODSS

As a result of reduced emissions because of the Montreal Protocol and its Amendments and Adjustments, mixing ratios for most ODSs have stopped increasing near the Earth's surface. The response to the Protocol, however, is reflected in quite different observed behaviour for different substances. By 2003, the mixing ratios for CFC-12 were close to their peak, CFC-11 had clearly decreased, while methyl chloroform had dropped by 80% from its maximum. Halons and HCFCs are among the few ODSs whose mixing ratios were still increasing in 2000. Halons contain bromine, which is on average 40 to 50 times more efficient on a per-atom basis at destroying stratospheric ozone than chlorine.

However, growth rates for most halons have steadily decreased during recent years. Furthermore, increases in tropospheric bromine from halons have been offset by the decline observed for methyl bromide since 1998. Atmospheric amounts of HCFCs continue to increase because of their use as CFC substitutes. Chlorine from HCFCs has increased at a fairly constant rate of 10 ppt Cl yr–1 since 1996, although HCFCs accounted for only 5% of all chlorine from long-lived gases in the atmosphere by 2000. The ODPs of the most abundant HCFCs are only about 5–10% of those of the CFCs. The behaviour of CFC-11 is representative of the behaviour of CFCs in general, and the behaviour of HCFC-

22 and HFC-134a is representative of HCFCs in general. The most rapid growth rates of CFC-11 occurred in the 1970s and 1980s. The largest emissions were in the NH, and concentrations in the SH lagged behind those in the NH, consistent with an inter-hemispheric mixing time scale of about 1 to 2 years.

In recent years, following the implementation of the Montreal Protocol, the observed growth rate has declined, concentrations appear to be at their peak and, as emissions have declined, the inter-hemispheric gradient has almost disappeared. In contrast, HCFC-22 and HFC-134a concentrations are still growing rapidly and there is a marked inter-hemispheric gradient. Ground-based observations suggest that by 2003 the cumulative totals of both chlorine- and bromine-containing gases regulated by the Montreal Protocol were decreasing in the lower atmosphere. Although tropospheric chlorine levels peaked in the early 1990s and have since declined, atmospheric bromine began decreasing in 1998.

The net effect of changes in the abundance of both chlorine and bromine on the total ozone-depleting halogens in the stratosphere is estimated roughly by calculating the equivalent effective stratospheric chlorine. The calculation of EESC includes consideration of the total amount of chlorine and bromine accounted for by long-lived halocarbons, how rapidly these halocarbons degrade and release their halogen in the stratosphere and a nominal lag time of three years to allow for transport from the troposphere into the stratosphere. The tropospheric observational data suggest that EESC peaked in the mid-1990s and has been decreasing at a mean rate of 22 ppt yr^{-1} over the past eight years.

Direct stratospheric measurements show that stratospheric chlorine reached a broad plateau after 1996, characterized by variability.

OBSERVED CHANGES IN STRATOSPHERIC AEROSOLS, WATER VAPOUR, METHANE AND NITROUS OXIDE

In addition to ODSs, stratospheric ozone is influenced by the abundance of stratospheric aerosols, water vapour, methane and nitrous oxide. During the past three decades, aerosol loading in the stratosphere has primarily reflected the effects of a few volcanic eruptions that inject aerosol and its gaseous precursors into the stratosphere. The most noteworthy of these eruptions are El Chichón and Pinatubo. The 1991 Pinatubo eruption likely had the largest impact of any event in the 20th century, producing about 30 Tg of aerosol that persisted into at least the late 1990s. Current aerosol loading, which is at the lowest observed levels, is less than 0.5 Tg, so the Pinatubo event represents nearly a factor of 100 enhancement relative to non-volcanic levels. The source of the non-volcanic stratospheric aerosol is primarily carbonyl sulfide and there is general agreement between the aerosols estimated by modelling the transformation of observed OCS to sulphate aerosols and observed aerosols.

However, there is a significant dearth of SO_2 measurements, and the role of tropospheric SO_2 in the stratospheric aerosol budget–while significant–remains a matter of some uncertainty. Because of the high variability of stratospheric aerosol loading it is difficult to detect trends in the non-volcanic aerosol component. Trends derived from the late 1970s to the current period are likely to encompass a value of zero. The recent Stratospheric Processes And their Role in Climate Assessment of Upper Tropospheric and Stratospheric Water Vapour provided an extensive review of data sources and quality for stratospheric water vapour, together with detailed analyses of observed seasonal and interannual variability.

The longest continuous reliable data set is at a single location is based on balloon-borne frost-point hygrometer measurements and dates back to 1980. Over the period 1980–2000, a statistically significant positive trend of approximately 1% yr–1 is observed at all levels between about 15 and 26 km in altitude. However, although a linear trend can be fitted to these data, there is a high degree of variability in the infrequent sampling, and the increases are neither continuous nor steady. Long-term increases in stratospheric water vapour are also inferred from a number of other ground-based, balloon, aircraft and satellite data sets spanning approximately 1980–2000, although the time records are short and the sampling uncertainty is high in many cases.

Global stratospheric water vapour measurements have been made by the Halogen Occultation Experiment satellite instrument for more than a decade. Interannual changes in water vapour derived from HALOE data show excellent agreement with Polar Ozone and Aerosol Measurement satellite data, and also exhibit strong coherence with tropical tropopause temperature changes. The Boulder and HALOE data show reasonable agreement for the early part of the record but there is an offset after 1997, with the Boulder data showing higher values than HALOE measurements. As a result, linear trends derived from these two data sets for the period 1992–2004 show very different results. The reason for the differences between the balloon and satellite data is unclear at present, but the discrepancy calls into question interpretation of water vapour trends derived from short or infrequently sampled data records.

It will be important to reconcile these differences because these data sets are the two longest and most continuous data records available for stratospheric water vapour. It is a challenge to explain the magnitude of the water vapour increases seen in the Boulder frost-point data. Somewhat less than half the observed increase of about 10% per decade can be explained as a result of increasing tropospheric methane. The remaining increase could be reconciled with a warming of the tropical tropopause of approximately 1 K per decade, assuming that air entered the stratosphere with water vapour in equilibrium with ice. However, observations suggest that the tropical tropopause has cooled slightly for this time period, by approximately 0.5 K per decade, and risen slightly in altitude by about 20 m per decade. Although regional-scale processes may also influence stratospheric water vapour, there is no evidence for increases in tropopause temperature in these regions either. From this perspective, the extent of the decadal water vapour increases inferred from the Boulder measurements is inconsistent with the observed tropical tropopause cooling, and this inconsistency limits confidence in predicting the future evolution of stratospheric water vapour.

The atmospheric abundance of methane has increased by a factor of about 2.5 since the pre-industrial era. Measurements of methane from a global monitoring network showed increasing values through the 1990s, but approximately constant values during 1999–2002. Changes in stratospheric methane have been monitored on a global scale using HALOE satellite measurements since 1991. The HALOE data show increases in lower stratospheric methane during 1992–1997 that are in reasonable agreement with tropospheric increases during this time.

In the upper stratosphere the HALOE data show an overall decrease in methane since 1991, which is likely attributable to a combination of chemical and dynamical influences. Measurements of tropospheric N_2O show a consistent increase of about 3% per decade. Because tropospheric air is transported into the stratosphere, these positive N_2O trends produce increases in stratospheric reactive nitrogen which plays a key role in ozone photochemistry. Measurements of stratospheric column NO_2 from the SH and the NH mid-latitudes show long-term increases of approximately 6% per decade and these are reconciled with the N_2O changes by considering effects of changing levels of stratospheric ozone, water vapour and halogens.

OBSERVED TEMPERATURE CHANGES IN THE STRATOSPHERE

There is strong evidence of a large and significant cooling in most of the stratosphere since 1980. Recent updates to the observed changes in stratospheric temperature and to the understanding of those changes have been presented of WMO. Current long-term monitoring of stratospheric temperature relies on satellite instruments and radiosonde analyses. The Microwave Sounding Unit and Stratospheric Sounding Unit instruments record temperatures in several 10–15 km thick layers between 17 and 50 km in altitude. Radiosonde trend analyses are available up to altitudes of roughly 25 km. Determining accurate trends with these data sets is difficult. In particular the radiosonde coverage is not global and suffers from data quality concerns, whereas the satellite trend data is a result of merging data sets from several different instruments. It reveals a strong imprint of 1 to 2 years of warming following the volcanic eruptions of Agung, El Chichón and Mt. Pinatubo. When these years are excluded from long-term trend analyses, a significant global cooling is seen in both the radiosonde and the satellite records over the last few decades. This cooling is significant at all levels of the stratosphere except the 30 km level in the SSU record. The largest global cooling is found in the upper stratosphere, where it is fairly uniform in time at a rate of about 2 K per decade. In the lower stratosphere this long-term global cooling manifests itself as more of a step-like change following the volcanic warming events. The cooling also varies with latitude.

The lower stratosphere extratropics show a cooling of 0.4–0.8 K per decade in both hemispheres, which remains roughly constant throughout the year. At high latitudes most of the cooling, up to 2 K per decade for both hemispheres, occurs during spring. Much recent progress has been made in modelling these temperature trends. Models range from one-dimensional fixed dynamical heating rate calculations to three-dimensional coupled chemistry-climate models; many of their findings are presented later in this stage. FDH is a simple way of determining the radiative response to an imposed change whilst 'fixing' the background dynamical heating to its climatological value; this makes the calculation of temperature change simpler, as only a radiation model needs to be used. Changes in dynamical circulations simulated by models can lead to effects over latitudinal bands, which vary between models.

However, dynamics cannot easily produce a global mean temperature change. Because of this, global mean temperature is radiatively controlled and provides an important focus for attribution. For the global mean in the upper stratosphere the models suggest roughly equal contributions to the cooling from ozone decreases and carbon dioxide increases. In the global mean mid-stratosphere there appears to be some discrepancy between the SSU and modelled trends near 30 km: models predict a definite radiative cooling from carbon dioxide at these altitudes, which is not evident in the SSU record. In the lower stratosphere the cooling from carbon dioxide is much smaller than higher in the stratosphere. Although ozone depletion probably accounts for up to half of the observed cooling trend, it appears that a significant cooling from another mechanism may be needed to account for the rest of the observed cooling.

One possible cause of this extra cooling could be stratospheric water vapour increases. However, stratospheric water vapour changes are currently too uncertain to pinpoint their precise role. Tropospheric ozone increases have probably contributed slightly to lower stratospheric cooling by reducing upwelling thermal radiation; one estimate suggests a cooling of 0.05 K per decade at 50 hPa and a total cooling of up to 0.5 K over the last century. The springtime cooling in the Antarctic lower stratosphere is almost certainly nearly all caused by stratospheric ozone depletion. However, the similar magnitude of cooling in the Arctic spring does not seem to be solely caused by ozone changes, which are much smaller than in the Antarctic; interannual variability may contribute substantially to the cooling observed in the Arctic wintertime. One mechanism for altering temperatures in the upper troposphere and lower stratosphere comes from the direct radiative effects of halocarbons. In contrast with the role of carbon dioxide, halocarbons can actually warm the upper tropospheric and lower stratospheric region.

In summary, stratospheric temperature changes over the past few decades are significant and there are clear quantifiable features of the contributions from ozone, carbon dioxide and volcanism in the past stratospheric temperature record. More definite attribution of the causes of these trends is limited by the short timeseries. Future increases of carbon dioxide can be expected to substantially cool the upper stratosphere.

However, this cooling could be partially offset by any future ozone increase. In the lower stratosphere, both ozone recovery and halocarbon increases would warm this region compared with the present. Any changes in stratospheric water vapour or changes in tropospheric conditions, such as high-cloud properties and tropospheric ozone, would also affect future temperatures in the lower stratosphere. Furthermore, circulation changes can affect temperatures over sub-global scales, especially at midand polar latitudes. Several studies have modelled parts of these expected temperature changes.

THE EFFECTS OF CLIMATE CHANGE ON ECOSYSTEMS

One way that scientists gain understanding of how global warming will affect ecosystems is to analyse the effects of past climate on paleoecosystems. A paleoecosystem is an ecosystem that existed in a former geologic time period. By relating vegetative cover to past climates, models can be developed. Once a reliable model is created, input variables such as CO_2 can be varied and the results analysed. An example of what the effect would be like on populations of Douglas fir in the northwestern United States if the CO_2 content in the atmosphere were double what it was before the Industrial Revolution was modeled by the U.S. Geological Survey and is shown in the illustration that follows. In a study conducted by the U.S. Global Change Research Programme in Washington, D.C., first in 2001, then updated in 2004, in trying to predict what the effects of future climate change would have on ecosystems, they concluded that"climate change has the potential to affect the structure, function, and regional distribution of ecosystems, and thereby affect the goods and services they provide." They based their conclusions on a modeling and analysis project they conducted called the Vegetation/Ecosystem Modeling and Analysis Project.

This project was used to generate future ecosystem scenarios for the conterminous United States based on model-simulated responses to both the Canadian and Hadley scenarios of climate change. The VEMAP was subsequently used in a validation exercise for a Dynamic Global Vegetation Model by Oregon State University and the U.S. Forest Service in 2008. Their MC1-DGVM was used as the input data in both the VEMAP and VINCERA. Their MC1 was run on both the VEMAP and VINCERA climate and soil input data to document how a change in the inputs can affect model outcome. The simulation results under the two sets of future climate scenarios were compared to see how different inputs can affect vegetation distribution and carbon budget projections.

The results indicated that"under all future scenarios, the interior west of the United States becomes woodier as warmer temperatures and available moisture allow trees to get established in grasslands areas. Concurrently, warmer and drier weather causes the eastern deciduous and mixed forests to shift to a more open canopy woodland or savannatype while boreal forests disappear almost entirely from the Great Lakes area by the end of the 21st century. While under VEMAP scenarios the model simulated large increases in carbon storage in a future woodier west, the drier VINCERA scenarios accounted for large carbon losses in the east and only moderate gains in the west. But under all future climate scenarios, the total area burned by wildfires increased." The similarities of the two models served to validate the VEMAP project.

The Hadley model was developed by the Hadley Centre for Climate Prediction and Research in England. Also referred to as the Met Office Hadley Centre for Climate Change, it is based at the headquarters of the Met Office in Exeter. It is the key institution in the United Kingdom for climate research. It is currently involved not only with understanding the physical, chemical, and biological processes within the climate system, but also with developing working models to explain current phenomena and to predict future climate change. It also monitors global and national climate variability and change and strives to determine the causes of the fluctuations. The Canadian climate model was developed by the Canadian Centre for Climate Modelling and Analysis. The CCCma is a division of the Climate Research Branch of Environment Canada based out of the University of Victoria, Victoria, British Columbia.

Its specific focus is on climate change and modeling. In the past nine years, the CCCma has produced three atmospheric and three atmospheric/oceanic general circulation models, making them one of the international leaders in climate change research.

What they found was that over the next few decades climate change in the United States will most likely lead to increased plant productivity as a result of increasing levels of CO_2 in the atmosphere. There will also be an increase in terrestrial carbon storage for many parts of the country, especially the areas that become warmer and wetter. The southeast will most likely see reduced productivity and, therefore, a decrease in carbon storage. By the end of the 21st century, many areas of the country will have experienced changes in the distribution of vegetation. Wetter areas will see the growth of more trees; drier areas will have drier soils that will cause forested areas to die off and be replaced by savanna/grassland ecosystems.

Modeling the vegetation evolution and adaptation is more difficult. The study focused on two time periods: 2025-2034 and 2090-2099. In the near term, biogeochemical changes are expected to dominate the ecological responses. Biogeochemical responses include changes based on the natural cycles of carbon, nutrients and water. The responses are affected by changing environmental conditions such as temperature, precipitation, solar radiation, soil texture, and atmospheric CO_2. It is these natural cycles that affect carbon capture by plants with photosynthesis, soil nitrogen processes, and water transfer. These biogeochemical factors are what influence the production of vegetation. In the results from the near-term biogeochemical model, the scientists concluded that there would be an increase in CO_2. They estimate that currently the average carbon storage rate is 66/Tg/yr.

The Hadley model predicted that carbon storage rates by 2025-2034 would increase to 117/Tg/yr. The Canadian model estimated CO_2 to increase 96 Tg/yr. The Canadian model projected that the southeastern ecosystems will lose carbon in the near term because they predict the climate there will become hot and dry. The biogeography models look at the changing landscape based on changes in CO2, evapotranspiration, vegetation establishment, and competition between species, growth rates, and life cycle/mortality rates. In this model, scientists at both the Hadley and Canadian Centre for Climate Change agree that vegetation will be able to freely move from one location to another.

Changes in vegetation distribution will vary from region to region as follows:

- *Northeast*: Forests remain the dominant natural vegetation, but forest mixes will change. There will also be some increase in savannas and wetlands.
- *Southeast*: Forests remain the dominant ecosystem, but mixes change. Savannas and grasslands encroach on forests, especially towards the end of the 21st century. Drought and wildfires contribute significantly to forest destruction.
- *Midwest*: Forests remain the dominant land cover, but changes in species type occur. There will be a modest expansion of savannas and grasslands.
- *Great Plains*: Slight increase in woody vegetation.
- *West*: The areas of desert ecosystems shrink, and forest ecosystems grow.
- *Northwest*: Forested areas grow slightly.

A separate study conducted by the Canadian government predicts the following ecosystem changes as a result of changing climate over the next 100 years:

- *Coastlines*:
  - Flooding and erosion in coastal regions
  - Sea-level rise

- *Forests*:
  - Increase in pests
  - Increased levels of drought and wildfire
- *Plants and animals*:
  - Warmer temperatures could make water supplies more scarce, having a negative impact on plants and animals, not giving them time to adjust.
- *Crops*:
  - In some areas, warmer climate may allow a three- to five-week extension of the frost-free season, which could benefit commercial agriculture.
  - In other areas, drier soils and lack of water will have a negative impact on agricultural productivity.
- *Wells*:
  - The quality and quantity of drinking water may be threatened by increasing drought.
- Harsh weather:
  - Winter storms, floods, drought, heat waves, and tornadoes could become more frequent and severe.
- *Fisheries*:
  - Populations and ranges of species sensitive to changes in water temperature will be negatively affected.
  - Salmon harvests will be lower in the Pacific.
  - Changes in ocean currents may have a negative impact on the fisheries in the Atlantic.
- *Lakes and rivers*:
  - Water levels will decline under the influence of drought, negatively affecting drinking water quality.
  - Use of lakes for transportation, recreation, and fishing, and the ability to generate electricity may be curtailed under droughtlike conditions.
  - Other areas that may have an increase in precipitation may experience flooding, rising sea levels, and severe storms.

In February 2005, the Met Office in Exeter, England, issued a report titled"Avoiding Dangerous Climate Change." The objective of the study was to determine what levels of CO_2 were considered the tipping point for dangerous climate change with harmful effects on ecosystems and what actions could be taken now to avoid such an outcome. In the report, then prime minister Tony Blair stated:"It is now plain that the emission of greenhouse gases... is causing global warming at a rate that is unsustainable." Environment Secretary Margaret Beckett stated,"The report's conclusions would be a shock to many people.The thing that is perhaps not so familiar to members of the public... is this notion that we could come to a tipping point where change could be irreversible. We're not talking about it happening over five minutes, of course, maybe over a thousand years, but it's the irreversibility that I think brings it home to people." The report, published by the British government, says there is only a small chance of greenhouse gas emissions being kept below"dangerous" levels. It warns that the Greenland ice sheet could melt, causing sea levels to rise by 23 feet over the next 1,000 years. It also warns that developing countries will be the hardest hit. The report also states, based on the vulnerability of many of the world's ecosystems, that the European Union (EU) has adopted a target of preventing an increase in global temperature of more than 3.3°F (2°C). Some believe even that may be too high. The report states:"Above two degrees, the risks increase very substantially," with"potentially large numbers of extinctions" and"major increases in hunger and water shortage risks... particularly in developing countries." In order to meet their goals, British scientists have advised that CO_2 levels should be stabilized at 450 parts per million (ppm) or below. Currently the atmosphere contains 380 ppm. In response to this, the British government's chief scientific adviser, Sir David King, said that was unlikely to happen. He stated,"We're going to be at 400 ppm in 10 years' time. I predict that without any delight in saying it." Myles Allen, an expert on atmospheric physics at Oxford University, said that:"Assessing a 'safe level' of carbon dioxide in the atmosphere was 'a bit like asking a doctor what's a safe number of cigarettes to smoke per day.'"

The report does conclude, however, that there are technological options available to reduce CO_2 emissions that will need to be used. The study also concluded that the biggest obstacles involved with using these new technologies, along with renewable resources of energy and"clean coal," are the current economic investments and traditionally strong bond to the oil industry, cultural attitudes that oppose change, and simple lack of awareness by many people. Various conservation organizations currently involved in the battle against global warming, such as the Union of Concerned Scientists (UCS), the Defenders of Wildlife, and the World Wildlife Fund (WWF), also support these ideas.

10

Forests and Natural Environment

WATER, FORESTS, AND MINERALS

Water, forests, and minerals are interdependent. They are physically linked, for instance, in that minerals are found in forests and watersheds. Without water, forests cannot grow. Forests regulate water flows and act as barriers to coastal and river floods. Forests also contribute to climatic stability through carbon sequestration, which then stabilizes water supply. Because of such close linkages, the exploitation of one can lead to the degradation of another.

Drawing examples from South and Southeast Asia, this chapter highlights the ways in which mining, logging, and exploitation of water resources interact with one another. Recognizing such linkages is integral to addressing mismanagement and unsustainable exploitation of natural resources, and the resulting environmental degradation, natural disasters, loss of livelihoods, political instability, and conflicts.

In Asia, in recent years, there has been growing awareness of the interdependence of key natural resources. At the policy level, a holistic approach—one that recognizes such linkages—is essential to the long-term sustainability of all the resources. The first examines how logging, mining, and building dams for hydropower contribute to the degradation of the resources being exploited as well as of other resources.

A holistic approach, such as differing definitions of forests, competing sector interests and jurisdictions, lack of communication among the different sectors, insufficient data on how the exploitation of one resource affects another, and a development framework based on economic indicators. Although mining, forestry, and water sectors are affected similarly in terms of policy, there are also significant differences.

The impacts of logging are increasingly recognized by governments and citizens, especially with the global attention to climate change and severe floods that have affected Asia's major cities in recent years. Policymakers have paid less attention to how mining pollutes water and destroys forests, perhaps because of the following factors: the profitability of mining that encourages governments to overlook environmental impacts, the difficulty of monitoring these impacts, and a pervasive lack of transparency in the sector.

RESOURCE EXPLOITATION AND THE NATURAL ENVIRONMENT

IMPACTS OF MINING

Mining has increased over the past decade in many mineral-rich countries in Asia, including Cambodia, Indonesia, the Philippines, and India, to meet the rise in global demand for coal, copper, and other minerals. Although mining is a significant source of revenue and economic growth for many countries, development of the mining sector is accompanied by serious social and environmental impacts, especially the degradation of forests and water resources.

Most mineral deposits lie beneath forests and in watersheds that support livelihoods. Current extraction methods, such as blasting and drilling, are crude, and no effort is made to rehabilitate areas once minerals are exhausted, because reclaiming the land would cost more than the value of the minerals. The negative environmental impacts of mining involve the mining process itself and related activities: the elimination of waste, transportation, and the processing of minerals.

Forest degradation and water pollution caused by mining not only affects the ecological system of an area, but the livelihoods of people who depend on these resources for sustenance. The three main methods of mining are damaging in different ways: open-pit mining for hard rock metals; quarrying for industrial building materials, such as sand and gravel; and leach mining in which chemicals are used to separate metal, such as gold, from the rest of the minerals.

The scale of impact may seem greatest with open-pit mines, where all vegetation and soil are removed, dynamite is used extensively, and indigenous communities are often displaced. But quarrying leaves trenches in rivers, alters the aquatic ecosystem, and is likely to affect urban people. Companies regularly operate sand mines near cities in order to reduce transportation costs for the construction material.

Leaching involves the use of chemicals hazardous to the health of living organisms. Sadly, mining companies have operated in developing Asian countries with little concern for environmental impacts, and governments have lacked the administrative and technical capacity, and the political will, to effectively regulate these operations. The problem is exacerbated by the fact that mining deals between governments and businesses lack transparency, and efforts to properly regulate mining activities have been neglected because of mining's profitability. Exhausted mine sites are commonly abandoned, and the environmental damage is almost always irreversible.

GLOBAL DRIVERS OF NATURAL RESOURCE DEPLETION

Steadily rising global demand for raw materials, industrial inputs, and energy have been the main drivers of the depletion and degradation of natural resources in the three regions. China's hyper-growth has made it the single largest importer of natural resource–based commodities, and India is fast catching up. More recently, the rapid growth in global demand for energy has created a new Hobson's choice for many of the three regions' governments that subsidize food and fuel consumption.

In several countries, efforts to reduce or reallocate fuel and food subsidies have been met by mass demonstrations and violence. Except during the Great Depression, trade in natural resources has been increasing since the industrial revolution and the advent of fast and reliable motorized cargo ships. In recent decades, several developments have created an almost exponential growth in demand and prices. One of the most important has been the liberalization of trade and the free flow of international capital beginning in the late 1980s, commonly associated with the so-called "Washington Consensus" of the International Monetary Fund, the World Bank, and the US Treasury Department. The lowering of trade and investment barriers has generally fostered increased GDP growth in the low- and middle-income countries, but also rapidly growing income inequality as globalization has penetrated the world's remote primary forests, mineral deposits, and previously untamed rivers. In the past several years, two more factors have also driven demand for natural resources and related industrial commodities. China's preparations for the 2008 Beijing Olympics, which began in 2003, created a major spike in global prices for construction-related commodities, such as timber, plywood, steel, and cement. The natural resource–based construction materials that China imported constituted a significant part of the US$50 to 60 billion cost of preparing for the Olympics.

The speculative international financial bubble that began in about 2000 (and is now rapidly deflating) also created unsustainable levels of demand for natural resources. From 2002 to 2007, prices of natural resource–based commodities, such as metal and energy, grew by as much as 350 percent.

As of late 2008, prices of energy and other industrial inputs have begun to fall, but global financial markets have lost 40 to 60 percent of their pre-crisis value, and most countries are struggling with a serious banking crisis and taking measures to stave off a deep economic recession. If there is any silver lining to the dark financial clouds, a short- to medium-term slowdown in the global demand for energy and natural resources could provide a breathing spell for the adoption of more sustainable resource policies.

THREAT TO REGIONAL STABILITY

Particularly because of its effect on the changing fortunes of nations, globalization is also a geopolitical phenomenon that poses a long-term challenge to interstate relations, and hence to peace and stability. Laurent Cohen-Tanugi, for example, argues that "economic globalization exists in a complex dialectic with the traditional geopolitics it has, ironically, helped to revive."

More simply put, globalization creates an ambivalent international environment that, on the one hand, encompasses both economic integration and regionalism, and on the other, fragmentation, conflict, and shifting centers of power. As a consequence, more traditional forms of power competition are carried out against a backdrop of emerging nontraditional security threats, such as terrorism, nuclear proliferation, and civil wars at the national, regional, and global levels. Resource scarcity can be a cause of conflict and the cause of further resource depletion.

CROSS-REGIONAL COMPARISONS

The Middle East, South Asia, and Southeast Asia have radically different resource endowments. The Middle East, rich in oil and gas, is poor in coal and metallic minerals. Its oilexporting countries, including Iran and Iraq, still account for nearly 30 percent of global crude oil production and hold about 55 percent of proven petroleum reserves. Less favorably, almost the entire Middle East has a serious water deficit and little exploitable forest land.

The Arabian Peninsula is one of the driest regions on earth. In some countries, energy resources are increasingly used for desalination, a process that includes huge capital investment and is causing degradation of shared water resources. The natural resource picture for South Asia is mixed. Much of the region, dependent on highly variable monsoon rains to replenish groundwater, frequently suffers both drought and severe flooding.

An ever-burgeoning population strains water supplies. The subcontinent has some of the world's largest rivers, including the Indus, Ganges, and Brahmaputra, which all originate in the Himalayas and Tibetan Plateau. All have been tapped for hydroelectricity production and irrigation. While minerals are an important resource, the region has few remaining stands of primary forest.

India and other South Asian countries are net importers of natural resources and related commodities, including timber, metals, coal, petroleum, and petroleum products. From 1999 to 2005, India's imports of wood products nearly doubled to about US$1 billion. Logs—primarily from Myanmar (Burma) and Malaysia—make up 88 percent of India's wood imports.

Despite their own significant domestic output, South Asian countries, including India, generally have emerged as major net importers of copper and other metallic minerals, as well as coal and coke (hard coal baked under high temperature to remove impurities and used in steel making). Southeast Asia is rich in forest, mineral, and water resources, but all are under severe pressure from population growth, the destruction of watersheds, and river pollution resulting from rapid urbanization, mining, and hydropower development.

Despite an alarming rate of deforestation, Southeast Asia still has some of the largest remaining stands of primary forest. Indonesia, Malaysia, and Myanmar (Burma) are all major log exporters. Strikingly, as of 2000, some 33 percent of Indonesia's land mass was in the hands of timber concessions. The region also has some of the world's largest deposits of coal, tin, nickel, copper, and gold. Indonesia's exports of coal and metal ores grew by about 18 percent per year from 2000 to 2007, and accounted for close to 8 percent of government revenues and 4 percent of the GDP.

The vast and highly controversial operations of subsidiaries of Freeport McMoRan Copper & Gold (a US company) in Indonesia's province of West Papua (formerly Irian Jaya) include the world's largest gold mine and third largest copper mine. Since major policy reforms in the 1990s, both China, and to a lesser extent India, have become voracious consumers of raw materials from Southeast Asia.

Both are major importers of timber and wood products from the neighboring region. Despite their own significant production of coal and minerals, China and India together are the most important markets for coal, coke, copper, and other minerals, as well as rubber and edible oils from Southeast Asia.

They both have also tapped into the hydroelectric potential of their less developed neighbors. On a cross-regional basis, the *causes* and *problems* of inadequate and/or unsustainably developed natural resources tend to be more similar than different. The main differences stem from the different resource endowments and/or stages of development. Historically, resource-rich countries exploit those resources as the foundation for development and to strengthen the state. Forms of government and ideology have remarkably little influence on natural resource exploitation, except in the few countries with functioning democratic institutions and welldeveloped civil societies.

Regardless of professed ideology, most political systems have pursued the same policies toward resource exploitation: reducing direct costs by creating economies of scale. While the locus of resource policy formulation varies from country to country, long-entrenched bureaucracies dominate the execution phase. In remote areas, where natural resources are found, career officials of relevant ministries and police are the primary face of government.

The most important differences in outcomes have to do with governmental competence and capacity, the means of valuating resources, the extent of corruption—which is commonly widespread—and the political balance between vested interests (both state and private) and the interest of those who occupy the land whose resources are to be exploited. Somewhat ironically, countries dominated or strongly influenced by the military often appear the least alert to the national security implications of their natural resource policies.

The worst case is Myanmar (Burma), whose military regime is all but a kleptocracy. Natural resources also tend not to be well managed in countries where the armed forces have a widespread role in maintaining political order or are substantially supported by allocations of natural resource tracts rather than by the national budget. These include Pakistan, Indonesia (but less so since the withdrawal of the military from politics after the fall of Suharto), Laos, Cambodia, and Vietnam.

In Laos, three military-run companies control the allocation of timber rights to commercial loggers, many of whom are from China, Thailand, and Vietnam. By falsifying chain of custody certifications that the timber has been cut sustainably, corrupt military and civilian officials engage in wholesale cutting of assigned tracts, carry out illegal cutting in adjacent forests, and illegally export logs to neighboring countries.

WATER POLLUTION AND CONSUMPTION

Mineral reserves are often found in areas that are either near the origins or in the catchments of rivers. Mining significantly damages rivers and water supplies in a number of ways. Heavy metal pollution occurs when some metals (*i.e.*, arsenic, cobalt, copper, lead, and silver) found in excavated rock or exposed in an underground mine come in contact with water. Pollution in the processing stage occurs when chemical agents, such as the cyanide used to separate the target mineral from the ore spill, leak or leach from the mine site into nearby waters.

Mining companies may also purposely discharge tailings (waste of extracted ore) into waterways, particularly rivers and oceans, to cut production costs. Pollution and sedimentation of rivers and oceans caused by mining can threaten riverine and marine biodiversity, the health of nearby communities, and further impoverish those communities highly dependent on water. Tailings can affect the local availability of water as it breaches groundwater. For example, **40** percent of the captive limestone mines in India have breached the groundwater in their regions.

Run-off from deforested slopes makes rivers heavy with silt and more prone to flooding. Mining near river basins, especially in underground coal mines, involves additional risks of accidents from inundation. Toxic cyanide and mercury used in Cambodia, Myanmar (Burma), and the Philippines for processing gold have polluted surface and groundwater resources. The untrained use of these chemicals introduces a high risk of damage to the environment and can seriously affect the health of the community and the ecosystem.

The *Cambodia Daily* reported in late 2003 that cyanide pollution at Phnom Chi resulted in the death of mass quantities of fish and cattle, and the eruption of human illness along the Porong River. Small-scale gold mining in the Philippines has been the target of strong opposition in recent years because of its environmental and social side effects, the most serious of which is mercury pollution. The consumption of large amounts of water by mining and mineral processing adds stresses to water supply.

Seventy-seven million tons of water used to extract iron ore in India between 2005 and 2006 would have been able to meet the daily water needs of more than 3 million people. At the Neyveli lignite mines in Tamil Nadu, 40 million liters of water is pumped out and wasted every day. In the numerous mining areas of India, people are experiencing water scarcity and pollution from mining. Local communities in the Philippines fear that pollution and siltation of rivers caused by mining may deplete water resources, reducing rice production and fisheries.

IMPACTS OF LOGGING

Illegal logging, plantation development (especially palm oil), and mining have caused significant forest loss in many Asian countries. There has been increased awareness in recent years among policymakers, farmers, and the general public about the impact deforestation has on water, especially in exacerbating floods. According to a Filipino forestry expert, most forest areas in the Philippines are in watersheds, and there is growing awareness among people in these areas of the strong link between water and forests. They are also coming to understand the impact of upland human activity in the lowlands.

Forests and Water

Forests contribute to the constancy of water supply, reducing the risks of drought as well as floods. Forests help maintain a constant flow of water toward other ecosystems and urban centers. Natural resource experts in Thailand and the Philippines are concerned that there has been a decline in water supply during the dry season, and more severe floods during the rainy season.

They attribute this phenomenon partly to deforestation and the impacts of climate change. Environment experts, government officials, and the general public in many countries, including Indonesia, Malaysia, and the Philippines, have recognized that deforestation can increase the risk of disastrous floods.

Deforestation increases flooding in rainy periods because the soil cannot retain the water as it does when it is covered by forest. Every year, thousands of people in Bangladesh are displaced by river erosion from logging activities. Because the country is so densely populated, people are often pushed to live in unsafe coastal areas. An increasing incidence of disastrous floods has prompted governments in the Philippines, Pakistan, and elsewhere to ban logging altogether.

As Shahbaz and Suleri's paper on forestry in Pakistan points out, the federal government imposed a complete ban on logging in 1993, a year after the vanished forests in the northern watersheds were blamed for one of the worst floods in the country's history. In recent years, in many countries in Asia, governments and civil society have pursued reforestation initiatives, not only to mitigate the effects of climate change, but also to prevent disastrous floods.

The devastating impact of the tsunami that hit Asia in December 2004 has also helped build momentum for reforestation efforts. Wetlands International, for example, has been encouraging the people living in the Mahakam Delta of East Kalimantan, Indonesia, to look after the mangrove forests that have been increasingly converted to housing communities, because they are the "last line of defense" from floods.

In the Philippines, rehabilitating forests is high on the agenda, as heavy flooding and landslides that occur almost annually are often attributed to deforestation. Logging also affects water resources. Timber processing operations, like mineral processing, consume water and pollute rivers and streams.

ROLE OF WATER RESOURCES

Water is a resource that seems to bear the largest burden from the exploitation of other natural resources, including forests and minerals. Mining pollutes the water, and logging degrades water supply by denuding the forests that stabilize the watershed. This is of particular concern as water is increasingly emerging as a scarce commodity, fueled by population pressures, intensive irrigation, and erratic weather patterns brought on by global warming.

Construction of hydropower dams can also destroy forests. Major hydropower dam projects in several Southeast Asian countries have been preceded by devastating logging operations in prospective inundation zones and in their vicinity for road construction. These projects have also resulted in deforestation elsewhere, as the local communities, displaced by the dams, have had to clear forests in other areas so they can settle, grow crops, and build houses.

POLICY ISSUES

Definitional Issues Forests

Effective forest management is impeded by a fundamental problem: the absence of a clear and consistent definition of what constitutes a forest. As a result, much of the research on deforestation and proposed remedies is based on flawed or questionable data, which greatly complicates efforts to promote more effective resource management. Compiling accurate global statistics on the dynamics of deforestation and reforestation is also a challenge. Definitions of a forest vary across countries and organizations because of the diversity and abundance of the world's forests, differences in culture and forest use, and the stage of a society's development. The definition of a forest can broadly fit into one of three categories: an administrative or legal unit (lands that are legally proclaimed "forests"), land cover (the extent to which the land is covered by trees), and land use (how the land is used).

In Pakistan, as noted in Shahbaz and Suleri's forestry paper, "forest areas" are lands that are under the administrative control of the provincial Forest Department. This may include areas that do not have a single tree, while other areas with significant tree cover may not be called "forests" because they are outside the department's jurisdiction. According to a definition reached by the United Nation's Food and Agriculture Organization in 2000, "Forests are lands of more than 0.5 hectares, with a tree canopy cover of more than 10 percent, which are not primarily under agricultural or urban land use."

Some countries in the region are now seriously addressing this definitional problem. India, for example, long debated the definition of a forest before reaching an agreement last year on specific definitions for different types of forests. However, many issues remain to be negotiated and rules are required before this change can take effect.

Coastal Zones

In some countries, the definition and/or the property right of a coastal zone is unclear. This leads to inadequate zoning and other land use controls that contribute to ineffective management of mangroves. In Asian countries, there is also a great deal of variation in the national laws for integrated coastal management and in how they are enforced.

Obstacles to a Holistic Approach

Significant benefits can be gained by adopting a holistic approach in the management of natural resources and the environment. Farmers and local communities directly experience the impact that the exploitation of one resource has on another. NGOs are raising awareness and engaging in projects to address this issue, and natural resource experts are stressing the importance of taking an integrative approach. Policymakers are also recognizing the interdependence of the resources, as they too have been feeling the impact of the increasing scarcity of resources and of extreme weather: devastating droughts, floods, soil erosion, cyclones, and polluted waters. However, adopting a holistic approach to correct these problems in most, if not all, the developing countries in Asia is impeded by the following six obstacles.

Competing Sector Interests and Jurisdictions

Because forests, water, and mineral resources coexist and the exploitation of one resource harms another, different stakeholders fight over competing sector interests, including the various government ministries and businesses exploiting the resources. The clash over how to manage forest resources in Indonesia is a classic case of sectoral and jurisdictional conflict.

In Indonesia, many government institutions—namely, the Departments of Forestry, Finance, Industry and Trade, Mining and Energy, and the Ministries of Environment, Labour, and Resettlement—have direct or indirect interests in the management of forest resources. Each has different interests and views on how to manage the country's forests. The most important is the Department of Forestry.

According to the 1967 Forestry Law, all forests are under its control. But the mining sector has long fought with the forest department for mining rights on land within its jurisdiction. For years, BHPBilliton and other multinational companies put pressures on the Indonesian government to amend Forestry Law 41/1999, which banned operation of mines in protected forests.

As a result of such pressures, an amendment to the law was passed in 2004, allowing 13 mining companies that had acquired a mining contract before the enactment of the 1999 Forestry Law to resume open-pit mining in protected forests. In India, the Indian Bureau of Mines clears mine plans and closure plans, and oversees monitoring and regulation under the Mineral Conservation and Development Rules, which includes air pollution and discharge of toxic liquids.

Although the bureau is clearly responsible for environmental issues, it has no power to oversee environmental impact assessments and management plans, which are cleared by the MoEF. State pollution control boards (SPCBs) provide consent to establish and operate mines and monitor water and air pollution, much like the Indian Bureau of Mines but under different legislation: the 1974 Water (Prevention and Control of Pollution) Act and the 1984 Air (Prevention and Control of Pollution) Act.

According to a Bangladeshi mining expert, the main problem in his country is a lack of coordination among different government departments, resulting in mismanagement of projects, with only some of the projects stringently following regulations. In Sri Lanka, the Coast Conservation Act governs the coastal zone, but mangroves are governed by the Forest Act, which causes conflicts related to jurisdiction and poor enforcement of both laws.

Not only are there uncertainties and lack of coordination for resource management among the various central government institutions, but central and regional governments often clash over the issuance of logging or mining licenses. The issue is especially problematic in countries such as Indonesia that have recently undergone decentralization.

Vague Laws and Lack of Capacity

Vague laws and lack of capacity also play significant roles in natural resource management. The laws are particularly ambiguous in the mining sector. In India, the environmental law, especially tailored for mining under the **1998** Mineral Conservation and Development Rules, is full of indefinite statements, which leaves plenty of room for interpretation by miners. For example, the rules say that land should be restored to its original use "as far as possible."

They also say that trees logged to make way for mining have to be replaced by double their numbers but there is no mention of the type of afforestation to be done. Institutions responsible for managing the resources not only compete or are confused by overlapping jurisdictions, but are often too weak to effectively carry out their designated roles. According to the Centre for Science and Environment in India, the pollution control boards of mineral-rich states—Jharkhand, Orissa, and Chattisgarh—do not have the capacity to regulate mines.

Lack of Cross-Sectoral Discussions and Policies

Another obstacle is that various governmental institutions and technical experts working on different types of resources do not communicate enough with each other. As a result, the issues that affect all natural resources are not reflected holistically in policies and laws. In India, there is currently little research on how mining affects rivers and groundwater, and no legislation exists to regulate the effect of mining on water resources. There is a lack of transparency, especially in the mining sector, where much of the decisions are made behind closed doors, by the government and businesses.

Lack of Data

Although there is a general recognition that deforestation destabilizes water supply and that mining is destructive to forests and water resources, there is still insufficient information and statistics showing these relationships. There is a lack of data, for instance, on how degradation of one watershed affects flood risks in the Philippines. In India, although the civil society has been active in shedding light on the negative effects of sand mining, there has been little research on how mining affects rivers and groundwater in the country. The research on such linkages has not been carried out in most countries in the region due to lack of capacity or sufficient interest, or lack of communication between the scientists/ researchers and the general public and policymakers. Governments and businesses may also purposely withhold information that may reflect negatively on them or conflict with their development strategies. Lack of sufficient information hinders the ability of local communities to reject proposals from industries for natural resource exploitation that would have a negative impact on them.

Development Framework Based on Economic Indicators

The fact that countries put economic growth over protecting the environment is probably the most fundamental cause of resource degradation and social conflict. According to an Indonesian environmental expert, environmentalists recognize the linkages among the various natural resources, but difficulties arise in policymaking because development in Indonesia is based on economic indicators rather than an ecological or sustainable development framework.

Thus, water is seen as an economic commodity for agriculture, consumption, and industrial purposes, and timber as a concession to forest-based industries, but the impact of deforestation on the water supply is disregarded. A Filipino forestry expert laments that most of his country's policies are driven by political pressures rather than scientific imperatives, and science in his country has not played a significant role in designing reforestation projects.

A natural resource economist in Thailand believes there is not enough attention paid to the fact that the royalties from river sand mining in Thailand are too low because they do not include the costs of erosion—the externality that it creates. Despite the significant damage it causes to the environment and natural resources, mining has been promoted by many countries in Asia, including Bangladesh, India, Indonesia, and the Philippines, because it is highly profitable. In the interest of generating revenue, environmental regulations and monitoring have taken a backseat.

Efforts to eliminate illegal, small-scale mining and properly regulate large international companies have been largely unsuccessful. Governments of developing countries woo foreign investments by circumventing their own laws intended to protect the environment and human rights, and setting standards below international practice. The Indonesian government has given permission for submarine tailings disposal, while the United States and Canada have not because of its damage to the environment.

Dangers of Policies Backfiring

It should be recognized that simply banning logging or promoting reforestation activities will not immediately solve flooding or climate change issues. Instead, a logging ban might initially contribute to livelihood insecurity in communities that rely on the resources, and encourage illegal logging, or the exploitation of timber resources in neighboring countries. In 2004, when a major typhoon and massive landslides claimed the lives of thousands of people in Luzon, Philippines, all logging activities were suspended throughout the country in response to pressure from civil society.

A Filipino forestry expert observed that the government is too indiscriminate when reacting to disasters of this magnitude, and that banning logging in all forests in 2004 punished those who actually practice sustainable forestry management. The expert also added that there are no typhoons in Mindanao, but timber harvesting, the sole source of livelihood to some communities in the region, was taken away.

Moreover, laws prohibiting exploitation within a country have not prevented it in less affluent neighbors. For example, Thai businesses sought timber resources in Cambodia and Laos after logging was banned in their own country. Timber has been smuggled from Afghanistan into Pakistan since the latter banned commercial logging in 1993.

Finally, the establishment of national parks, such as those in Pakistan and Indonesia, for the protection of forests and biodiversity, might lead to increased degradation of natural resources because the local communities, which no longer have the sense of land ownership, are not as invested in conserving the resources.

COMMUNITY PROTESTS AGAINST MINING COMPANIES

Civil society and local communities in Asia have been actively protesting mining activities that cause water pollution and illnesses. The American-owned Newmont mining company attracted national attention in July 2004, when the people of Buyat Bay, off Sulawesi Island in Indonesia, demanded compensation for the disposal of tons of tailings containing high levels of mercury and arsenic pollutants into the ocean between 1994 and 2002.

The 20-month trial began after local villagers filed a US$543 million lawsuit against Newmont in August 2004, contending that tailings from its gold mine had caused serious illnesses and the ruin of their fishing income.

Initial assessment by the World Health Organization and the Indonesian Ministry of Environment found the water to be unpolluted, but a subsequent ministry study found that arsenic levels were 100 times higher at the wastedumping site than in other parts of the bay. The district court that cleared the case in December 2007 claimed that it could not prove the company polluted the bay, and the mine waste did not breach local or international safety standards.

More recently, in August 2008, the chief of a Papuan tribe in Mimika Barat Jauh, Kaimana district, West Papua, urged the government to take action against US mining company Freeport, and demanded compensation for contaminated rivers from mine tailings. The chief claimed that as a result of Freeport's mining operations, local village communities were facing water shortages, and that crocodiles and tortoises of Teluk Etna were on the brink of extinction.

In 2006, a toxic tailings spill at an Australian-run mine on Rapu-Rapu Island, Philippines, spurred mining opponents, including the Catholic Church, nationalistic politicians, and NGOs to demand a total ban on mining activities in the country. An independent commission report urged the company to stop all mining in the area, revoked its mining rights, and forced the company to pay damages. Philippine President Gloria Arroyo declined the commission's request to ban all mining activities, but promised to carefully study the report's recommendations, including a review of the 1995 Mining Act's provisions for foreign participation and management.

DEFORESTATION

Mining is one of the major causes of deforestation and forest degradation, as commercially valuable minerals are often found in the ground beneath forests. Large-scale, open-pit mining operations can result in significant deforestation through forest clearing in order to access mineral deposits and to open remote forest areas for miners. Infrastructure built for transient mine workers—roads, tunnels, and dams—also has an impact.

Large quantities of timber are often used as supports in mine shafts and in tunnels, in the case of underground mines, and as fuel for operating mines. This link is a major concern for India, where forest lands have been increasingly designated as mines in the last decade. Between 1980 and 1997, the Ministry of Environment and Forest (MoEF) only granted forest clearances for 317 mines, which destroyed a total of 34,527 hectares of forest area.

By contrast, from 1998 to 2005, the ministry cleared 881 mining projects in forest areas, destroying 60,476 hectares of forest area. As Chandra Bhushan of the Centre for Science and Environment in India observes, "This is an extremely worrying trend, since the mining industry has just begun expanding: with the new mining policy on the anvil, which is likely to give a significant boost to the industry, much more forest land is likely to be diverted for mining.

RIVER SAND MINING

Minerals and water resources are intimately linked in river sand mining. Extensive river sand mining substantially alters the physical, chemical, and biological environments of rivers. Mining and dredging activities, uncontrolled dumping of overburden, and chemical spills reduce water quality and poison aquatic life. Indiscriminate sand mining can make an area prone to flash floods. Sand mining has also been blamed for water shortage, as it deepens the riverbed and depletes groundwater.

Bank erosion and channel deepening from sand mining can also undermine nearby bridges and other engineering structures. With rapid urbanization and growth in the housing and infrastructure sector all over Asia, the demand for sand and gravel has grown significantly over the past few years. Most of this demand is being met by rampant, often illegal mining of riverbeds. In July **2008**, a professor observed the serious impact of sand quarrying in the Amaravathy River in Chettipalayam, India: "Amaravathy used to provide drinking water for the entire Karur town but now it is not even able to supply water to the villages on its bank."

In the Palakkad district, the diminishing stock of sand and increasing environmental concern forced the state government to ban sand mining in the district. But illegal mining still takes place, and some have been selling river sand mixed with cheap sea sand, which causes structural failures in buildings. Kelani River, one of the major rivers in Sri Lanka, has also been badly affected by sand mining in recent years.

Overmining has caused problems such as salination of Colombo's drinking water and the collapse of the river bank. However, it is difficult to ban sand mining in the river because many people living nearby depend on it as a source of living. Sand mining has considerable implications for food security. It not only leads to water shortages affecting agriculture, but many farmers, seeing the high profits to be made, have abandoned their traditional ways of rice farming for it. Some in Bantul, Yogyakarta, have realized that sand mining is more profitable and less labour intensive than rice production.

Mining activities are threatening the forests of other Asian countries as well. In October 2008, the Indonesian Mining Advocacy Network (JATAM) and WALHI-Friends of the Earth Indonesia, protested BHP-Billiton's destruction of protected forest in Central Kalimantan for coal mining and in Gag Island, West Papua, for nickel mining. From 1999 to 2004, BHP-Billiton, along with other multinational mining companies, put pressure on the Indonesian government to amend Forestry Law 41/1999, which prohibits open-pit mining in protected forests.

Despite public protests across the country, a new law was issued in 2004 that permitted companies holding mining contracts issued before 1999 to operate in protected forests, including BHP-Billiton and PT Gag. In Central Kalimantan, BHPBilliton's coal mine affects 65,858 hectares of protected forest covering the upper reaches of the area's main rivers.

Forest degradation due to mining projects has other effects: it causes a decline in the natural environment's productivity, and it renders the local community more vulnerable to fatal floods, landslides, adverse climatic shocks, and other natural disasters.

Some governments in the region have begun taking action against the devastating impacts of sand mining. In November 2008, Chief Justice Sarath Silva voiced his concern over illegal sand mining in Sri Lanka. Together with Environment Minister Champika Ranawaka, he recently participated in a progrmme to promote awareness of the serious consequences of illegal sand mining.

The government of Andhra Pradesh, India, has taken measures to control illegal and excessive sand mining by seizing vehicles that have been ferrying sand. Sand mining became a transboundary issue in February 2007, when the previous usual supplier, Indonesia, abruptly banned its export of sand to Singapore, citing the impact of a recent Singaporean construction boom. Mari Pangestu, Indonesia's Trade Minister, banned the exports, saying it was necessary to protect the environment and maintain her country's maritime borders, threatened by rampant sand removal. Smugglers can ship as much as 10,000 cubic meters of sand from coastal areas in a day. Indonesia's Maritime

Affairs Minister Freddy Numberi claimed that the ban was also aimed at pressing Singapore into signing a long-stalled extradition treaty. Singapore's Ministry of National Development expressed disappointment not only with the decision itself, but with Jakarta's rejection of Singapore's offer to help resolve Indonesia's environmental concerns. Not all sand mining has lasting effects if done responsibly; sand mining is nonchemical and no blasting is required.

However, there is still a need to increase awareness about the environmental impacts of increasing scarcity of sand and escalating construction costs. Research institutions in Asia are working to develop alternatives to river sand. The faculty of civil engineering of the Government College of Engineering in India has developed a technology to produce a cubical-shaped sand with excellent binding strength from hard granite.

NATURAL RESOURCE EXPLOITATION AND WIDENING INCOME GAPS

The assumption of state control over land and natural resources, and the poor governance of those resources, have become the two greatest threats to human security and livelihoods. Despite considerable efforts by the multilateral development banks (MDBs), bilateral aid donors, and governments themselves, efforts to involve affected communities in decisions about the development of natural resources have largely failed in South and Southeast Asia.

The reasons are not difficult to understand. Especially in relatively remote regions, someone nearly always has some kind of claim to land, fisheries, and minerals, either by right of customary use or legal title. These rights are almost always overridden by government and private sector development imperatives that involve interests that are far more powerful than local rights.

Typically, local communities are consulted long after the key decisions have been made. Since the colonial era, the main objective of the state and forestry departments has been to alienate the people from their land. "Timber mafias" operate in concert with corrupt officials. Community forestry experiments have generally failed because the process remains under the control of state bureaucracies, and because governments are more interested in managing forests for revenue than providing livelihoods.

Establishing trust between local communities and the state is impossible when the forest department police officer is the face of government. Ironically, deforestation often rises when governments alienate people from their lands by establishing state forests and other protected areas. Where governance is weak, those who have lost their lands have little hesitation about "illegally" continuing to utilize resources while assuming no responsibility for maintaining sustainable conservation practices and stewardship.

The situation is largely the same in regard to dam construction and the awarding of mining concessions. The people who will lose their land and livelihoods are, at best, represented perfunctorily in so-called "stakeholder" consultations that are dominated by government agencies and developers. The decisions were made long before—the only question is how much environmental damage will be mitigated, if at all, and how much effort and expense will be devoted to relocation and the provision of alternative livelihoods.

Mining may present the most difficult problem because mitigation is normally expensive and/or impractical, and the cost of returning land to its original condition is so high as to make it uneconomical. Although large commercial mining operations are often carried out with technically sophisticated machinery, mining remains crude by the standards of any other industry. Because of mechanization, the industry employs relatively few workers, and local people do not normally have sufficient skills for employment.

Mining is particularly prone to social conflict because, while big operators are given concessions by the government, numerous small, "artisanal" operators may already be working the sites, often backed by financing from criminal syndicates, which purchase and transport the output to the global market.

URBAN WINNERS AND RURAL LOSERS

The inescapable reality is that economic development in poor but resource-rich countries involves the exploitation of rural-based resources, such as timber, minerals, and water, to serve the interests of the politically important urban areas. Pinkaew Laungaramsri underscores the inevitability of this process when he says, "Commodifying nature goes hand in hand with the growth of urban middle-class society and its increasingly intense lifestyle in big cities."

Despite the rationale that the exploitation of natural resources will give governments more money for anti-poverty progrmmes, rural villagers displaced by hydropower projects and other large-scale uses of natural resources are inevitably net losers. However poor their existing subsistence livelihoods, they almost always end up with insufficient compensation and lands that are less productive than those from which they were evicted.

Fishermen are relocated to areas without fisheries, forest people must leave entirely or take insecure jobs as plantation workers, and farmers often have to learn to grow new crops on less fertile land. Rural dwellers in South and Southeast Asia, as in other less developed regions, bear the brunt of natural resource–based development. Because of the failure of most environmental mitigation and relocation efforts to date, displaced people who already live a subsistence existence more likely than not will become part of the swelling ranks of the urban poor, often homeless and jobless.

The 60 to 80 percent of the population in developing countries that live in poor rural areas is effectively subsidizing the lifestyles of the 20 to 40 percent that live in cities. This is a global phenomenon, but the tragedy in South and Southeast Asia, as in Latin America, is that most countries are not failed states in which competition for control of valuable resources is a source of civil war and external conflict. Not yet, at least. Rather, these destructive decisions about resource exploitation are made by at least nominally responsible political leaders and bureaucracies.

DEVELOPMENT FAILURES AND THEIR CAUSES

In theory, developing countries should be making a transition from commodities production to higher value added production, as has occurred in the more advanced developing countries such as Thailand, Malaysia, and China, but those countries remain the exceptions. In fact, for a variety of reasons, the poorest countries have done little to reduce their dependence on natural resource exports, whether legal or illegal, and instead have traded forests for oil palm, rubber, and acacia plantations that provide comparatively little employment.

In general, the poorest and most politically marginalized citizens are the primary victims. In reality, natural resource exploitation is an essential, but not a sufficient, component of broad-based and stable economic development. Even more important are education, technical training, and other forms of human capacity building. In tropical Southeast Asia, only Thailand, Malaysia, and Singapore thus far have made the transition to middle-income status.

As shown by the Asian financial crisis, even these gains tend to be subject to reversals, especially since only tiny Singapore, which has no significant natural resources, has made the necessary investments in human capital to sustain its gains in the face of rising competition from China.

One obstacle to balanced and sustainable development is that the industrializing countries of South and Southeast Asia often become the exploiters of resources in neighboring countries with poor governance and systemic corruption.

Laws against timber cutting in natural forests that were enacted in China, India, Thailand, Malaysia, Pakistan, and other countries with few remaining primary forest resources have not stopped deforestation. Instead, companies from these countries have moved into neighboring countries that are weakly or corruptly governed and still have exploitable timber. Chinese timber and plantation companies appear to operate with impunity in Cambodia, Laos, and Myanmar (Burma).

Malaysian companies are particularly active in Myanmar (Burma) and Indonesia. Pakistan imports timber from Afghanistan. As the process of deforestation proceeds, Western and Asian companies are looking further afield, to formerly remote parts of the islands of Borneo (Indonesia, Brunei, and Malaysia) and New Guinea (Indonesia and Papua New Guinea) that are the "new frontier" for timber and extractive industries in Asia and Australasia.

Freshwater and ocean fisheries are likewise under relentless pressure from soaring demand and diminishing supplies. Even water becomes a global commodity when the dry countries of the Gulf invest in the production of wheat, rice, and other food crops in developing countries with water and irrigable land. For instance, Saudi Arabian investment in wheat production in Pakistan and imports of rice from South and Southeast Asia represent a transfer of water from water-rich areas to dry regions. Sadly, resentment at what is seen as Western arrogance and a perceived double standard regarding environmental concerns has caused some countries, especially in Southeast Asia, to reject opportunities to leapfrog over the most environmentally destructive forms of resource-based development. The argument is frequently heard and read that Western admonitions regarding unsustainable development are self-serving and in conflict with their rich countries' own early development phases.

Many also argue that the West is ignoring the pull factor in natural resource exploitation, *i.e.*, that the very developed countries that are most concerned about climate change and other aspects of natural resource degradation are the main customers for these resources and products. This gives rise to considerable self-defeating cynicism. For instance, some political leaders and opinion makers in resource-rich developing countries argue that it is the Western countries that overconsume the world's resources and have the biggest carbon footprint. The developed countries, the argument continues, should cut back their own CO_2 emissions and be willing to compensate the developing countries for practicing sustainability. Others talk about a related but different kind of developed country "footprint," that of the multinational companies that lay claim to natural resources, both financing and otherwise carrying out mining, for example, as well as transporting and marketing the resultant ore or coal. This perspective is understandable but counterproductive. Moreover, some of the argument ignores important differences in the development experiences of the rich countries compared to the less developed ones. Some of the biggest per capita carbon consumers, such as the United States, Canada, and Australia, are lightly populated in comparison with the developing countries of the Middle East and Asia, which have little or no room for expansion into new frontiers. Water in most of the three regions is already oversubscribed and polluted. Forests in the northern and southern latitudes regenerate more quickly and easily than tropical forests. In other words, once the finite natural resources of the three regions are depleted, the adjustment will be especially wrenching.

A WINDOW TO THE WORLD FOR THE CUSTODIANS OF THE FOREST

Isang Bagsak South-East Asia is an enabling programme in more ways than one. The implementation of this networking programme, through its basic strategy of participatory development communication (PDC), is giving rise to experiences and insights on involving the community in development work through communication. More significantly, this involvement hinges on the use of information and communication technology – a 21st-century tool whose applications for development are currently gaining ground in developing countries.

The programme is being implemented in the Philippines by the College of Development Communication (CDC) of the University of the Philippines at Los Baños and its partner in the first cycle, Tanggapang Panligal Para sa Katutubong Pilipino, or the Legal Assistance Centre for Indigenous Filipinos (PANLIPI). The decision regarding the Philippine partner for the Isang Bagsak learning network was not an easy one to make. There were many mainstream organizations with vast experience in natural resource management. Resource constraints also figured prominently on the choice of partner.

In the end, the CDC posed one question that made all the difference: who would benefit most from the Isang Bagsak experience? The following rhetorical question also tipped the balance in favour of PANLIPI: why not share the learning from Isang Bagsak directly with the indigenous peoples, the custodians of Philippine forests?

According to the Indigenous Peoples' Rights Act of the Philippines, indigenous peoples refer to a group of people or homogeneous societies identified by self-ascription or ascription by others who have continuously lived as organized communities in a communally bounded and defined territory and who have, under claims of ownership, since time immemorial, occupied, possessed and utilized such territories sharing common bonds of language, customs, tradition and other distinctive cultural traits. Indigenous peoples have, through political, social and cultural inroads of non-indigenous religions and cultures, become culturally differentiated from the majority of Filipinos. Indigenous peoples will include people who are regarded as indigenous on account of their descent from populations who occupied the country before colonization and who retain some or all of their own socio-cultural or political institutions.

A SHOWCASE OF SYNERGY

PANLIPI is an organization of lawyers and advocates of indigenous peoples' concerns. Established in 1985, PANLIPI primarily aims to assist the indigenous peoples in their struggle for the recognition of their rights to their ancestral domains, culture and traditions, and other basic rights. The end goal is to empower the indigenous peoples to the fullest so that they can actively participate in every aspect of Philippine society. PANLIPI's mandate includes the provision of development legal assistance; legal education and outreach; institutional capacity-building; ancestral domains delineation; and resource management planning.

The networking among Isang Bagsak, CDC and PANLIPI may be viewed as a synergistic one right from the start. From PANLIPI's viewpoint, the promotion of the rights of indigenous peoples to their ancestral domains and the issue of natural resource management go hand in hand. The indigenous peoples are, after all, the best caretakers or stewards of the natural resources mainly because their socio-economic system is sustainable and not destructive.

For its part, CDC found in PANLIPI a partner who could best make use of the very tenets of development communication – that is, the use of communication towards improved socio-economic growth of a community that makes for social equity and the larger unfolding of individual potential. Through Isang Bagsak, the indigenous peoples of the Philippines are given their window to the world of learning and are sharing their own knowledge, insights and experiences with members of the network from various parts of the globe. Likewise, the other members of the network are allowed access to the rich heritage of peoples hitherto unwired to the Isang Bagsak network.

LESSONS LEARNED

Working with PANLIPI is, indeed, a work in progress that is bringing forth a lot of valuable insights for researchers and development workers in general. The introductory stage of the partnership was already fraught with lessons, foremost of which were the ethical considerations. For one, is it ethical to bring in the indigenous peoples to a programme that is quite alien to them in terms of the tools (information technology) to be used? Second, will they agree to become part of the undertaking, and how do we engage their participation?

MUTUAL RESPECT AND TRUST

There was conscious effort on the part of the CDC to show respect and to gain the indigenous peoples' trust at the start of the project. Before any agreement was inked, the project team, with PANLIPI's help, made a point of observing two things:

- understanding the cultural context; and
- securing the indigenous peoples' free and prior informed consent.

Henceforth, a series of consultations with the indigenous peoples who would be involved in the learning network was undertaken. The CDC team, together with the PANLIPI lawyers, literally sailed, crossed rivers and hiked through mountains to dialogue with the elders of the cultural communities. This was part of securing their free and prior informed consent, in keeping with the PANLIPI protocol and the nature of participatory development communication. These are two critically important values that many development workers seem to be taking for granted, but which the Isang Bagsak experience is bringing to the fore in this undertaking. Genuine dialogue between and among the CDC, PANLIPI and the indigenous peoples paved the way for the participation of people from four areas: Mangyans from Mindoro; Tagbanuas from Palawan; Kankanaeys from the Cordilleras; and Aetas from Zambales. Iterative consultations leading to consensus-building were held. First, the CDC team conferred with the PANLIPI staff and lawyers assigned to the areas. Once PANLIPI's cooperation was ensured, dialogues were held with indigenous elders and leaders, and eventually with the members of the community.

The series of dialogues highlighted the following points:

- Knowledge sharing: Isang Bagsak offered an opportunity for the PANLIPI network and the indigenous peoples in the four areas to not only learn from others and from one another about natural resource management and PDC, but to contribute their own knowledge to the network.
- Assurance that the indigenous knowledge systems (IKS) would be respected: the indigenous peoples voiced a strong concern about the possible piracy of their traditional knowledge, as was their experience in the past. They were wary of outsiders 'stealing' their indigenous knowledge in the guise of development work. Prior experience of being exploited by researchers from outside (foreign and local) was a big stumbling block to securing the indigenous peoples' free and prior informed consent. Team members took pains to explain that nothing that did not come from the people, through their representatives in the programme, would be posted on the website. This was, in essence, a manifestation of respect for intellectual property rights.

RELEVANCE TO INDIGENOUS PEOPLES' CONCERNS

The indigenous peoples' decision to participate in Isang Bagsak rested largely on the realization that the skills they would develop would be useful to the concerns that they were advocating, such as the Ancestral Domains Sustainable Development and Protection Plan and other important provisions of the newly passed Indigenous Peoples' Rights Act. PANLIPI participants saw the Isang Bagsak experience as an excellent opportunity to interact with professionals from other countries. Through the sharing of experiences, one important insight surfaced: the affirmation that all along, PANLIPI and the indigenous peoples have been practising the principles of development communication, in general, and those of PDC, in particular. The training that the PANLIPI participants have undergone in the course of their participation in Isang Bagsak has also afforded them the opportunity to write and communicate their ideas clearly.

INFORMATION TECHNOLOGY FOR INDIGENOUS PEOPLES

Technology could be an awesome development for indigenous peoples, only very few of whom have been initially exposed to computers, much less to the internet. Part of their initial reluctance to be part of Isang Bagsak was apprehension about the technology itself. Connectivity/accessibility also loomed as a potential problem.

To address these concerns, PANLIPI, for its part, has begun sponsoring a series of training sessions on information technology for the indigenous leaders involved in Isang Bagsak.

CDC's information technology team trained Iraya Mangyan leaders in Mindoro during February 2004. A user-friendly sourcebook on using the computer and the internet was developed by the CDC especially for this training. The sourcebook, in turn, would be useful in future information technology training for indigenous peoples. In addition, PANLIPI has been very supportive in enabling the indigenous peoples to have access to computers and the internet.

CDC and PANLIPI team members agree that one of Isang Bagsak's 'byproducts' is the gradual mainstreaming of the indigenous peoples into the digital age.

PDC THROUGH THE NET: TEAMWORK AT ITS BEST

By the time the project was in full swing, the PANLIPI–Isang Bagsak network had 45 members, 9 each from the 5 indigenous groups. A network within a network was created by the partners. As the saying goes: *t*ogether, *e*veryone *a*chieves *m*ore.

The members were selected by the indigenous communities themselves on the basis of their knowledge and skills on natural resource management issues. Other criteria were also included in the list of qualifications.

Prospective team members had to be knowledgeable about natural resource management and indigenous knowledge systems and practices. They had to be capable of facilitating community participation. Finally, they were also expected to understand development work and to be knowledgeable and skilled in promoting a rights-based approach to development.

The resulting team was dubbed *Limang Tagupak*, a Mangyan term representing victory. The name also signifies a five-point star, representing the configuration of the five-team network at the PANLIPI level. The star is also symbolic of a guide.

The sharing and learning process on PDC in natural resource management was carried out in the following manner:

- Local teams undergo orientation and training on Isang Bagsak.
- Local teams participate in the forum:
    - o The CDC posts theme questions.
    - o PANLIPI National Capital Region translates theme questions and sends them to local teams.
    - o Local teams discuss the theme questions, formulate answers and send them to PANLIPI National Capital Region.

  - o PANLIPI National Capital Region translates answers and posts reply.
  - o PANLIPI National Capital Region downloads and translates country comments and sends to local teams.
- Local teams discuss country comments and make their own reflections, which they send to PANLIPI National Capital Region.
- PANLIPI National Capital Region translates and posts reflections.
- The CDC resource person synthesizes.
- PANLIPI National Capital Region translates the synthesis and sends it to the local teams.
- Local teams discuss the synthesis and extract the learning.

The process appears to be a tedious one. True enough, a number of problems hampered the implementation of Isang Bagsak at the PANLIPI level. For one, translations into English took up quite some time; hence, delays in posting were inevitable. Delays were also due to the team members' tight work schedules or their preoccupation with economic activities, leaving them less time to participate in the forum. Access to the internet (lack of equipment, telephone lines, internet terminals/connection) constituted the hardware part of the limitations. Financial costs of consultations and regular meetings, as well as internet rental, posed very real problems as well. So did forces of nature, such as bad weather preventing team members from meeting.

PANLIPI's response to the resource-related problems mirrored its commitment to the partnership. PANLIPI put up counterpart funds for training indigenous peoples on basic computer use and internet navigation, as well as hardware for internet accessibility. It forged cooperative undertakings with non-governmental organizations (NGOs) and other friends to boost its investment in equipment and internet access. To facilitate meetings among the team members, PANLIPI made arrangements for Isang Bagsak activities to coincide with regular meetings of indigenous peoples' organizations.

The foregoing are just four of the insights arising from several months' implementation of Isang Bagsak in the Philippines. Many more could be drawn from the researchers' and the indigenous peoples' experiences as they go through the cycle in the remaining months. Isang Bagsak South-East Asia is a work in progress.

FROM RESOURCE-POOR USERS TO NATURAL RESOURCE MANAGERS

Conservation efforts often fail to take into consideration that alternatives have to be available if local populations are to change their natural resource utilization patterns. Along the Cimanuk River in West Java, efforts to prohibit agriculture on the raised riverbanks to curb seasonal flooding irremediably failed until viable economic alternatives were introduced. Through participatory development communication, the dialogue between researchers, decision-makers and small and landless farmers made it possible for sheep raising to flourish as an alternative to the practice of agriculture on the riverbanks that was making villages more vulnerable to annual flooding.

Fifteen years ago, as we were preparing a baseline study to explore the development options that could improve the living conditions of small farmers and landless farm labourers living along the Cimanuk River in the district of Majalengka, West Java, we wondered whether those resource-poor farmers would be able to participate in managing a strip of public land stretched out along the river. We also wondered whether they would be able to derive benefits from such an undertaking.

At that time, the main concern of the local government and of the communities living along the river was flood control. Every rainy season, heavy downpours filled up the river and flooded the villages. To control this annual flooding, the Department of Public Works bought a strip of land along the river that goes through the district's villages and small towns, and raised the riverbanks to a certain height. Hence, the annual flooding was controlled and the communities were freed from the yearly threat of natural disaster.

However, this flood prevention technique created another problem for the small and landless farmers who had been farming on the riversides long before the banks were raised. According to the new government rules and regulations, farming activities were forbidden on the public land along the river due to the risks of destabilization of the land structure and hindrance of the waterways.

In spite of the government rules and regulations, the small and landless farmers pursued their farming activities as usual. On their part, officials of the Department of Public Works continued to enforce the regulations by eradicating the crops that could possibly weaken the raised riverbank's land structure.

Looking for Alternatives

After several years of playing cat and mouse and tired of conflicting with the local communities, the Department of Public Works finally proposed a win–win solution to the farmers. The latter were finally allowed to farm along the riversides, provided that they grew grass at least 1m away from the river's brim and on both sides of the raised bank. However, they were not allowed to grow bananas or any other type of land crops that could destabilize the land structure.

To further motivate the farmers to grow grass, the government officials promised to supply them with sheep. However, this promise did not materialize very well. The farmers were disappointed with the few low-quality sheep delivered to them and the abundant grass was wasted. At that critical time, a baseline study was undertaken in order to understand the availability of local natural resources such as land, crops, animals and water, as well as the limitations, problems and opportunities associated with their use and development. Moreover, the study also aimed at understanding the traditional systems and at assessing indigenous capacity. The study concluded that there was potential for involving the small farmers in managing the public land located along the river and the irrigation channels, and for deriving benefits from it, provided that a funding body be put in place.

A follow-up pilot study was then undertaken. Essentially, the challenge consisted in helping farmers to solve their problems and benefit from the abundant grass, while ensuring the conservation of the riverbank. After discussing the best-suited approach, our team decided to use participatory development communication. Thus, linkages were established between university researchers, local government, the Department of Public Works, the local livestock extension service, the village authorities and local farm communities.

USING APPROPRIATE COMMUNICATION TOOLS

Prior to the fieldwork, communication materials such as sound-slide shows, posters and even a comic book were prepared and tested with farmers and extension workers. Development communication graduate students were involved in producing and testing these materials, either through their message design course assignments or through their theses research.

Topics such as the social and economic potential of sheep raising in the district of Majalengka were specifically addressed with local decision-makers and livestock extension workers. Other topics concerning various technical aspects of sheep production, marketing and rural family budget management were developed for extension workers, farmers and their wives.

The fieldwork started once the communication materials were ready. In the first quarter of 1992, a memorandum of understanding was officially signed between the government of the district of Majalengka and the Faculty of Animal Science at Bogor Agricultural University.

During the ceremony, a sound-slide show regarding the social and economic potentials of sheep raising in the district of Majalengka was shown to the government officials and to the local House of Representatives, prior to signing the agreement. This created enthusiasm among the officials attending the ceremony. Moreover, the local media covered the event, which was broadcast on the evening news by the regional television station. The next step consisted of promoting capacity-building for the local livestock extension personnel and farmer leaders. Once a month, researchers conducted an on-site training meeting for the livestock extension workers and other officials of the district livestock service. The technical aspects of sheep production, including housing, feeding, breeding and sheep healthcare, were addressed. Two sound-slide programmes about sheep production were shown and discussed with the extension workers. A pre-test and a post-test on the training subjects were always administered to the extension workers prior to and after the slide show in order to build up the agents' interest and to focus their attention on the subjects. Moreover, the tests were also useful in indicating whether the sound-slide shows had been effective in improving the agents' knowledge on the topics that were being discussed.

Following the agents' training, a short meeting was conducted for farmer leaders of the four pilot sites. On that occasion, they were informed that they were entitled to receiving an in-kind loan in the form of sheep. If they agreed, they would have to return a certain number of the sheep offspring to the project for a period of four years.

After the meeting, participants were invited to visit a senior farmer leader in the village of Balida to witness how he collected forage. One month later, they joined an excursion to the district of Garut to visit a sheep-raising initiative. The Garut sheep were particularly well known for their fast growth and prolificacy.

Trying out New Techniques

Upon their return, every farmer leader was asked to build a good sheep house capable of housing two adult rams and ten ewes, and to prepare a plot to grow improved varieties of grass and legumes to feed the sheep for demonstration purposes.

Intensive supervision and backstopping activities were provided to the farmer leaders in the following month to allow them to get first-hand experience in raising the new sheep breed in their own environment. After one month, the farmer leaders had enough experience in taking care of the sheep and were able to provide information about the Garut sheep raising to other farmers in the villages.

Upon completion of the trial run among the farmer leaders, monthly evening training meetings were conducted for the rest of the farmers in the four research sites. The first training meeting was about building a good low-cost and healthy sheep house using local materials, while the second one focused on the appropriate feeding for this specific breed of sheep. During the meetings, all farmers were given pre-tests and post-tests in order to measure their knowledge regarding the training subjects prior to and after the audio-visual presentations.

About two weeks after the training sessions, the farmer leaders were asked to check whether all the farmers had made or had improved their sheep houses for the incoming new sheep. Farmers who had not done so were encouraged to rebuild or improve their sheep houses.

Two weeks after completing the necessary preparations, a farmer leader with extensive experience in sheep breeding was asked to join the researchers to select rams and ewes for 100 farmers in the four research sites.

The training meetings continued after the sheep had been distributed to the farmers. The topics addressed ranged from technical aspects of sheep raising to rural family budget management.

Meanwhile, the research assistant, together with the farmer leaders, visited every farmer once a week to check on the sheep's condition and health and to provide further backstopping activities to the farmers.

About one year after the distribution of the sheep, all farmer leaders were invited to a meeting in the district capital to share their experience in Garut sheep raising with farmers from other villages. This farmer-to-farmer communication turned out to be an excellent medium for raising other farmers' interest in Garut sheep production, better than anything that researchers or extension workers could have done.

Promising Results

The initiative had a first positive result when the head of the local livestock service, using the narrative progress report for the first year, was able to convince the local government to provide his office with extra funding

Towards the end of the second year, results began to be experienced at the grassroots level, when several farmers began to return the sheep offspring to the initiative. According to a previous agreement, the sheep offspring were to be split fifty–fifty.

The returned offspring were to be revolved to other farmers interested in sheep raising. This was an indication that the initiative was having a positive impact on the small farmers involved. Indeed, between September and December 1993, the sheep population increased by 159 per cent. One year later, the increase reached 271 per cent, while at the end of 1995, it had risen to 306 per cent.

The value of the participating farmers' assets also increased accordingly. In less than three years, the value of the sheep almost tripled – a positive contribution to the local economy. Moreover, the welfare of most farmers involved in the initiative improved significantly since the income generated by selling the sheep was worth more than 1.5 times the capital loaned to them.

In addition to the economic benefits, such as having more sheep in barns and gaining additional income, the initiative also brought about some social benefits to the villagers involved, who gained respect from their families and communities. One of the farmer leaders even managed to send his son to the Bogor Agricultural University, where he obtained his degree in Animal Science last year.

In May 1996, after about five years of continued fieldwork, the initiative officially came to an end. However, since the ewes continued to give birth and farmers kept on returning the sheep offspring, the revolving activities continued on their own.

A New Spur to the Initiative

In early 1999, following the severe Asian financial crises that badly hurt the Indonesian economy, the initiative obtained additional funding to buy 55 new sheep. These sheep were distributed to two other villages in the district of Majalengka.

Meanwhile, around the same time, one former project group in the village of Kadipaten received a 300 million Indonesian rupiah loan through the World Bank's social security network funds. The purpose of this loan was to help the local communities cope with the economic crisis. The farmer group used it to further develop their sheep raising ventures.

Eight years after the initiative terminated, we are still serving the farmers, though at a slower pace and on a smaller scale. To cover the operational costs of supervision and backstopping activities, fieldworkers have been allowed to sell the culled sheep.

Today, livestock production, including sheep production, is growing at a good pace in the district of Majalengka. Both public and private investments have been increasing steadily, especially since the 1998 economic crisis. Between 2002 and 2004, the government of Majalengka invested over 9 million rupiahs in livestock production.

Public funds have been channelled to farmers as loans through local banks, mainly for beef and cattle production. In addition, a big livestock marketing infrastructure has recently been built in the region. Livestock traders from east and central Java bring and sell their sheep and cattle in the newly operated livestock market to local and other traders and buyers from Bandung, Bogor and Jakarta.

In other words, the situation of livestock economics in the district of Majalengka has greatly evolved compared to the situation that prevailed in 1989, when the research team first arrived. We believe that over the last 15 years, our work has contributed to this situation by raising a flag to both public and private audiences, informing them about the existing potential in a remote district of West Java that was waiting to be explored and developed.

During recent times, when driving along the river and the irrigation channels in the former research sites, one can see that the irrigation channels are well maintained and that farmers periodically harvest the grass on the riverbanks. From a distance, one can see numerous small white plastic flags stuck orderly in the ground every 10m, on both sides of the irrigation channel. What do these flags mean? Are they a border line? Yes, they are. No one is now allowed to harvest the grass except for the farmer who sticks the flags on the ground. The grass now means something to the rural communities as it has played a key role in local economic development, while contributing to solving environmental problems.

In summary, through participatory development communication, researchers, extension workers, government officials and cooperating small farmers were able to work hand in hand to solve certain community problems. Through the introduction of sheep raising as an alternative economic activity, the small and landless farmers living in the farming communities along the sides of the Cimanuk River have been able to ensure the management of the meagre public natural resources in their environment.

Problems Encountered

This story would not be complete without describing some of the problems that the researchers encountered in the course of the initiative.

The first problem occurred during the early phase of the fieldwork, when we were trying to gain acceptance and support from the village heads. Most of the village heads in the research sites had shown great interest in the benefits that they could obtain from the activities that were to occur in their areas, and they were keen on having a cut.

Denying this opportunity to the village heads could have meant troubles ahead or, worse, it could have led to the failure of the initiative altogether. In contrast, letting them have a share of the benefits could possibly spare the researchers and the farmer leaders many feuds and tensions with the village officials.

For those reasons, the wisest choice at that time seemed to be to give them a small slice of the benefits. With that in mind, we included the village heads' relatives on the list of sheep recipients, provided that they agreed to abide by the project's policies and rules. It seemed to us that by doing this, we would save a lot of energy and avoid further conflict between farmer leaders and village heads. This compromise would also make it possible for us to use the village halls for training meetings.

As the activities unfolded, we found out that high mortality rates occurred among the project sheep raised by the village heads' relatives. Why? The answer was simple. They were not farmers and did not have the necessary skills to raise sheep. Moreover, they did not spend enough time on their tasks. Consequently, they failed in raising the sheep and also failed in returning the sheep offspring.

The second problem we encountered was the implementation of inappropriate feeding practices. Farmers were used to feeding natural grass to their animals and gave similar feeds to the new prolific sheep, without adding enough legumes in the rations. As a result, many animals suffered malnutrition.

Malnutrition brought about many miscarriages among the pregnant ewes and high mortality rates among newly born lambs. To solve these problems, the research team introduced urea molasses block as a protein and mineral supplement for the sheep.

The third problem was the lack of grass and forage supply during the dry season. This recurring problem was seasonal and continuously haunted the farmers year after year. At the beginning of the fieldwork, the research team, together with certain farmer leaders, initiated a demonstration plot to cultivate improved grass and legume varieties in the four research sites. However, very few farmers were willing to cultivate the grass and the legumes, arguing that the grass supplies had always been abundant. Why bother cultivating it?

Another difficulty encountered was that farmers did not respond well to the suggestion of silage-making as a way of preserving the grass, particularly during the rainy season, when there is a large supply of grass. They did not like its smell.

Lastly, after the project's end, when we had to reduce the frequency of our visits to farmers due to limited funds, many farmers did not keep their promise to return good-quality sheep offspring to the initiative. This situation badly affected the efforts to revolve the sheep offspring to other potential farmers. The sustainability question then came up. How much longer would the project last?

Despite these shortcomings, this initiative provided a number of learning opportunities. Looking back on the work accomplished, we can draw lessons and identify the most outstanding features that contributed to making it a success story.

First, conducting a comprehensive baseline study on the research sites was very useful in order to understand the situation, the problems and the opportunities related to local natural resource development, as well as the traditional systems and indigenous capacities of local communities. This also allowed for the development of a sound action plan to implement the initiative.

Second, the strong commitment of a funding partner to ensure adequate funding in order to implement the action plan and prepare the communication materials proved to be a key issue. Furthermore, support from local policy-makers, village authorities and farmer leaders also proved extremely important to ensure the successful implementation of the action plan.

The direct involvement of local extension workers and farmer leaders in disseminating the materials, in distributing the sheep and in providing backstopping activities also seems to have greatly contributed to building the necessary trust with local farmers. Moreover, holding monthly meetings proved to be a good way of getting feedback and monitoring activities, while providing further information to farmer groups about sheep raising and marketing their products.

Finally, holding dissemination seminars on the project's success, with the participation of farmer leaders, the academic community and policy-makers, was very important for the project's replication and expansion.

11

Community Forestry Environment

COLLABORATIVE MANAGEMENT OF COMMUNITY FORESTS

In response to a community forestry environment that is complex and rapidly changing, CIFOR began a multi-country research project in 2000, which aims to enhance forest management decision-making at the local level.

This research project, 'Adaptive and Collaborative Management of Community Forests' (ACM), explores the potential role of *collaboration* and *social learning* in forest management, including the role of criteria and indicators (C&I) as a tool within that process. Research hypotheses include that self- or collaborative monitoring systems can support communities in deepening their knowledge about local systems and impacts of management strategies, as well as creating and focusing dialogue between diverse stakeholders. The underlying assumption is that these changes can (in some conditions) support equity, effectiveness and adaptiveness in community forestry decision-making.

The objective of the current research at the meta level is to generate insights into three questions. Does *collaboration* among forest stakeholders, enhanced by conscious and deliberate *social learning* processes in forest management, lead both to improved human well-being and to the maintenance of forest cover and diversity? If this is so, under what conditions does it occur? And, what are the key strategies, approaches and tools to enable these processes? These research questions are rooted in the assumption that the challenge of incorporating multiple interests at multiple scales into participatory interventions has not yet been successfully met in NRM.

APPROACH

The current ACM research is rooted in a participatory action research (PAR) approach. In most communities involved, diverse local people and other relevant stakeholders jointly developed a set of agreed and easily understood C&I. The process provided an opportunity for communication and learning within and across the stakeholder groups, especially with regard to visions and goals. The C&I set also provided a framework for later monitoring and assessing of key factors and their direction of change. This monitoring process creates the opportunity to feed information and learning back into the community forest management system. It thus serves to guide future action, helping to increase the sustainability of the community's forest resources. It was initially the researchers who offered and provided the framework for the social learning process; since that time local users have begun to adapt and apply these processes themselves. The ultimate goal is to completely transfer these, including the necessary facilitation skills.

Community forestry systems are complex and dynamic settings with multiple stakeholders, overlapping and differing interests, capabilities, and a myriad of challenging livelihood activities and processes. In some countries, such as those in Zimbabwe, the action research focus on collaboration has included power relations and negotiations *between* local peoples and other stakeholders. In other sites, such as Nepal, the focus has been primarily on stakeholder relations and equity *within* the local forest user groups (FUGs). Researchers there have tried to understand the stakeholder diversity within the FUG – based on *overlapping* categories including gender, caste, ethnicity, wealth and geography – in terms of issues of equity, power and access to resources and decision-making.

The short-term outcomes of the self-monitoring processes and follow-up actions appear to be contributing to positive change in this area. In follow-up to the monitoring workshops, for example, some of the forest user groups are shifting their committee-based decision-making processes (which were generally dominated by the elite) towards hamlet and interest group-based processes, including building mechanisms for feedback to the committee. Especially given the linkages between hamlets, ethnicity and wealth in some of the FUGs, these changes have the potential to help address some long-standing local equity issues.

A PAR methodology in isolation would present challenges in terms of producing generalizable results (a CIFOR mandate); thus, to enable generalizability, the PAR is embedded in a larger multi-site framework of scientific analysis. Specifically, in all research sites, researchers have laid the foundations for the comparative research by conducting a series of background studies elaborating stakeholder relations, historical, biophysical and socioeconomic contexts and initial levels of adaptiveness and collaboration. The studies took a consultative form of participation, but allowed researchers the time to build relations and the groundwork with local stakeholders for the main PAR phase of the research. Additionally, processes and learning in all sites are regularly recorded by researchers in a framework and format that is comparable across countries and sites. Other methods are also being used to triangulate the results across sites, countries and regions, key among these being the use of multivariate analysis across all sites and the analysis of the outcomes of the participatory modelling in Zimbabwe, Indonesia and Cameroon. Whereas the multivariate analysis is expected to provide a quantified picture of key drivers

for the success or failure of adaptive collaborative management processes, the analysis of the simulation models is expected to provide insights into the causality of failure or success arising out of the structure and behaviour of these processes. Ultimately, these elements of the larger framework for analysis will enable greater depth of understanding within each site and highlight findings that emerge across varying community forestry conditions.

Reflections

This is an ambitious project with high local and research expectations, and as such it faces some significant challenges. At the meta level, two of the most critical challenges are those of working across so many diverse sites and countries, compounded by the limitation of a very tight three-year time frame. Key challenges to working at the community level include: complex and pervasive hierarchical local stakeholder relations; low social capital; unstable political climates; and geographical isolation. However, key strengths include that, on the whole, community stakeholders, district and national partners, and field researchers have a high level of commitment to exploring the process and seeking local benefits – both social and environmental. The PAR and collaborative approach to the research incurs time costs to researchers but is enabling lessons to be relatively rapidly shared and incorporated to the research as it progresses.

Past CIFOR C&I research fulfilled its intention of generating useful and valid insights for some national, regional and global stakeholders through traditional social and biophysical research. In that research context, relatively few benefits were intended to accrue, nor did accrue directly to the communities where the research took place. The current research is focused on community-level processes and makes a conscious effort to bring good science into a coherent, integrated framework with local learning and benefits. The outcomes are not yet assured – the approach is new and certainly bears some risk. But the indications so far are that, in the context of these issues at least, a synergy exists between functional and empowering participation that will be well worth the costs.

COMMUNITY-BASED NATURAL RESOURCE MANAGEMENT ORGANIZATION

A Comité de Investigación Agrícola Local (CIAL) or Local Agricultural Research Committee is a research service belonging to, and managed by, a rural community. The research group is made up of volunteer farmers, chosen for their aptitude in experimentation. The CIAL links farmer-researchers with formal research systems, increasing local capacity to exert demand on the formal system and to access potentially useful skills, information and research products.

A CIAT team developed the CIAL concept, which was first tested in Colombia. Five CIALs were formed in Cauca, Colombia, in 1990 and NGOs began their involvment in 1991. During 1995–1996, the CIAL project spread to other countries, involving rural communities, GOs, NGOs and universities. At present, over 250 CIALs are active in eight Latin American countries.

Experience has shown that the CIAL can benefit the wider community as well as individual CIAL members. Mature CIALs often launch small businesses, selling improved seed or other products or services. Many take on a broader role in the community, seeking access to credit and training, preparing and submitting proposals and acting as ambassadors in relationships with research and development (R&D) actors.

Approach

The main CIAL principles are:

- Building on experience and learning-by-doing to generate knowledge.
- Mutual respect and accountability and shared decision-making as the foundations of the relationship between the CIAL, the community and external actors.
- Risk shared by partners in the research process.
- Systematic comparison of alternatives for improving farming and natural resource management.
- Research products are public goods.

Each CIAL has at least four elected members (many expand beyond this) and a facilitator. The facilitator may be a trained agronomist from a supportive formal research centre or university, an extension service or an NGO, or a trained farmer, who has served on a CIAL. The facilitator plays a key role in developing the CIAL's competence in the research process, and provides feedback on farmers' priorities and research results to formal research and extension services.

The CIAL process cycles through the following stages:

- Training of facilitators and site selection by committed research organizations, extension services, NGOs, or community-based organizations.
- Mobilization though an initial community meeting and regular interaction between the CIAL and the community as the research process unfolds.
- Election, when the community chooses farmer researchers who are interested in experimenting and willing to provide a service to the community.
- Diagnosis, when the topic or question for experimentation is determined through a group process in an open community meeting.

- Planning, based on information from various sources and on a clear objective for each experiment.
- Experimentation, when the CIAL implements the planned experiment.
- Evaluation, when the CIAL meets with the facilitator to evaluate collected data, draw conclusions and present results to the community.
- Analysis, in which the question, 'What have we learned?' is especially important, ensuring learning from both the process and the results.
- Feedback, for which regular, open meetings are held in the community where the CIAL presents results and expenditures. This ensures that their research products become public goods. The facilitator is responsible for feedback to the formal research system.
- Regional or national forums where CIALs can exchange information and results are held annually.

Reflections

Ecological interactions and local knowledge gaps are not always considered explicitly in the CIAL process, and no formal mechanism is incorporated to develop an understanding of the agroecological principles and interactions that may underlie the issues being researched. Such knowledge may be essential for the design of meaningful experiments. Methods for analysing the scale of action needed for successful intervention are not sufficiently developed. This can be limiting if the research issues are related to pests, diseases and many NRM situations. Relationships with the community depend strongly on the quality of social capital. In areas where trust is low and association along non-kinship lines is rare, forming and/or sustaining CIALs may be difficult. However, many advantages accrue. Farmers systematically evaluate technological options such that minimal risk is incurred. They learn basic research principles that provide for developing a common language and two-way communication with research and extension professionals. In some conflict-ridden communities CIALs have provided an entry point for building social capital. Although community-based, the radius of CIAL influence can be increased by forming networks of experimenting communities and creating second-order associations.

A recent impact assessment of the CIAL movement revealed that:

- Many CIALs provide an effective research service to their communities; they report their results publicly and cite widespread testing of these by local farmers.
- In some communities where farmers are applying CIAL recommendations, yields of staple crops have almost doubled. Food security can be greatly improved in CIAL communities. The poorest benefit the most from the increased availability of food during times of scarcity.

- More farmers in CIAL communities are experimenting with soil conservation practices and conduct experiments with a far greater diversity of varieties and crops than in communities without CIALs.
- Innovations identified by the CIALs reach local farmers more rapidly than others and also spread to other communities both with and without CIALs.
- CIALs report positive changes in the attitudes of the R&D professionals working with them.
- Women and marginalized social groups gain social status and respect in their communities as a result of belonging to a CIAL.

The effect of the CIALs on their communities and on formal research services transcends dollars and cents. Nevertheless, CIAT estimates the return on the investment made in developing and applying the CIAL approach at 78 per cent.

The CIAL movement is still young, and its future evolution is uncertain. Properly managed, CIALs can deliver substantial growth and equity benefits. However, their effects on the sustainability of production are less predictable. By allowing adaptive research to be devolved to the farming community, CIALs cut the costs of formal research while increasing its impact.

AN INSTITUTIONAL INNOVATION WITH IMPLICATIONS FOR EXTENSION AND RESEARCH

Watershed degradation does not have to be an inevitable consequence of using sloping land for agriculture. Smallholders can farm and manage natural forest resources in a manner that is both productive and resource conserving. Awareness of this fact has focused attention on evolving approaches to watershed resource management that are demand-driven and community-based. In such approaches, those who occupy the land actively participate in managing and sustainably using their local watershed resources for multiple purposes. Watershed farming systems are enormously variable, and simple recipes do not solve their problems. Often, the issues need to be tackled at a larger scale than the individual household, cooperatively at the community level. In Asia, the role of local organizations in the management of forest and other common natural resources has received much attention. Joint forest management in India, forest users' groups in Nepal, and community-based forest management in the Philippines are notable examples. Similarly, local organizations may apply knowledge to solve problems in agriculture through improved land husbandry. In countries where power and fiscal responsibility are being decentralized, democracy is reaching the village level, and rural people are acquiring new leadership skills. These skills provide a basis for developing farmer-led organizations that can develop practical ways of achieving a more sustainable agriculture.

APPROACH

Landcare is a particularly noteworthy model for strengthening local initiatives to reverse land degradation. Through this approach, local communities organize efforts to solve agricultural and environmental problems in partnership with public institutions. Landcare groups are voluntary and self-governing. They engage local communities in a search for innovations that are suited to the diverse and complex environments of smallholder farming. They mobilize communities to address problems of water quality, forest and biodiversity protection, soil conservation and others at the landscape level.

The Landcare movement in the Philippines began in Claveria, Mindanao, in 1996. Now, about 200 village-based Landcare groups are working in Claveria and other municipalities in northern, central, southern and eastern Mindanao, with a membership of several thousand households. The groups have established more than 1500 conservation farms and more than 200 community and household nurseries that have produced hundreds of thousands of fruit and timber tree seedlings, all with local resources.

Conservation farming based on contour buffer strips has become popular as a result of collaboration between ICRAF and Landcare groups in the Philippines. With a view to diversifying farm enterprises, the groups have also established nurseries for new species of fruit and timber trees. At the community level, Landcare has proved itself a powerful force for creating initiatives that protect the whole watershed. Because the group members determine agendas, they have addressed a wide range of issues, including beef and dairy farming, cut-flower production and vegetable crop farming.

Landcare provides important opportunities for improving how farmer participatory research is done. Landcare groups can manage such research, enabling them to diversify their experimentation, ensuring a better understanding of the performance and recommendation domains of technical innovations and offering more effective and less expensive alternatives to technology-transfer approaches. The farmer field school approach for conservation farming is currently being explored as a means of initiating Landcare groups.

These groups exhibit some similar characteristics to the farmer field schools made popular in integrated pest management (IPM). Landcare groups, however, are more formalized and aim at a broader range of land degradation and sustainability issues. Some distinguishing features of Landcare groups are that:

- They develop their own agenda and tackle the range of sustainability issues considered important to the group.
- They tend to be based on neighbourhoods or small watersheds.
- The impetus for formation comes from the community, although explicit support from outside may be obtained.

- The momentum and ownership of the group's progrmme is with the community.

Reflections

The sustainability of the Landcare movement gives rise to three significant concerns. First, given its growing popularity, the movement runs the risk of 'projectizing', that is, attracting the support of projects that do not understand the concept, and that provide funds in a top-down, target-driven mode, defeating the whole basis of a farmer-led movement. Second is the issue of long-term sustainability. Networking and the stimulation from outside contacts are considered to be crucial for long-term success. This can be achieved through Landcare federations, as has evolved locally in Claveria, and through provincial and national federations, which are currently being explored in the Philippines. Third, group leadership is a time-consuming and exhausting task, particularly when undertaken on a voluntary basis. Landcare is still young in both the Philippines and Australia but leadership 'burnout' has already raised concerns.

The ICRAF analysis indicates some steps for further releasing the power of the Landcare concept. Public institutions and NGOs need to facilitate group formation and networking among groups, enabling them to grow, developing their managerial capabilities and enhancing their ability to capture new information from outside local communities. Such organizations can also provide leadership training to farmer-leaders, thus helping to ensure the sustainability of the Landcare groups. External financial assistance is also needed with an emphasis on the use of trust funds that enable farmer groups to compete for small grants to implement their own local Landcare projects. This approach has been remarkably successful in the Australian Landcare movement.Experience in the Philippines and Australia suggests that Landcare may provide an effective means of generating and sharing technical information, spreading the adoption of new practices, enhancing research and fostering farm and watershed planning processes. In the southern Philippines, Landcare groups are forming partnerships with local governments and technical research and extension agencies. Local governments are actively assisting the movement through financial and political support. This has attracted the attention of the national government, resulting in a national strategy of watershed management based on Landcare that will spread its principles and experiences to other parts of the Philippines.

MANAGEMENT OF PLANT GENETIC RESOURCES IN AGROECOSYSTEMS

In 1995, national partners in Burkina Faso, Ethiopia, Hungary, Mexico, Morocco, Nepal, Peru, Turkey and Vietnam together with IPGRI initiated a global project, 'Strengthening the scientific basis of *in situ* conservation of agricultural biodiversity'. The purpose is to strengthen the scientific basis, institutional linkages and policies that support the farmers' role in conservation and use of crop genetic diversity. Projects are implemented by linking existing national plant genetic resource (PGR) programmes with other partners such as universities, national institutes, agricultural extension workers, NGOs, community-based organizations (CBOs) and farmers. An expected result is the creation of a channel for input from these sources into national agricultural research agendas. A cross-section of case studies includes 'Creating a national framework' (Nepal), 'Understanding farmer preferences' (Morocco), 'Gendered participation in *in situ* conservation' (Mexico) and 'Participatory methods to add value to PGRs' (Nepal).

APPROACH

On-farm conservation of traditional crops is carried out by farming communities for farming communities. The primary task for those concerned with conservation and with the maintenance of traditional crop diversity *in situ* is to understand when, where and how this will happen, who will maintain the material and how those maintaining the material can benefit. Four areas of investigation were identified that set the necessary scientific agenda needed to support farmers and local communities in *in situ* conservation on farm:

1. What is the extent and distribution of the genetic diversity maintained by farmers over space and over time?
2. What are the processes used to maintain the genetic diversity on-farm?
3. Who maintains genetic diversity within farming communities (men, women, young, old, rich, poor, certain ethnic groups)?
4. What factors (market, non-market, social, environmental) influence farmer decisions on maintaining traditional varieties?

Formulating and answering these questions required a participatory approach at all stages of the process. Participatory methods, such as key informant interview, focus group discussions, spatial mapping and matrix ranking served to include farmers' knowledge on local social-cultural, economic and agroecological conditions, their crop and seed management practices and the characteristics and origins of their varieties into project data. Information from participatory research is complemented by household, market and seed system surveys, field trials on-station and on-farm and genetic diversity measurements in the field and in laboratories by economists, sociologists, ethnobotanists, agronomists, ecologists and populations geneticists, often in collaboration with farmers and extension workers. A preliminary 'exploratory approach' (not based on preliminary hypotheses) was used first, because it did not presuppose or assume the different categories or reasons underlying farmers' knowledge and it enables farmers to employ their own values and standards of measurement. Hypothesis testing followed, to answer specific questions that would support the scientific basis of on-farm conservation. The different sources and levels of information include the variety, crop, parcel or plot, household, village or community, landscape or region. Information is collected, disaggregated by gender and in some cases by wealth categories and ethnic groups.

Moreover, information from one aspect may be useful to answer more than one question. Understanding the relationship between what farmers recognize as or name a variety and the genetic distinctiveness of this unit is key to understanding the amount and distribution of crop genetic diversity managed by farmers. The information collected in only one year at the level of the household or farmer's plot may not be the appropriate scale for analysis or for agrobiodiversity conservation. Thus, information is analysed at different spatial and temporal scales.

Implementing the project involved developing multi-institutional, multidisciplinary collaboration at international, national and local levels. This included ensuring that trained national male and female personnel were available to carry out the work at central and local levels, and that the teams promoted equity at all project levels. In addition, because the project is largely community-based, time was devoted to building or creating rapport with the farmers in whose fields much of the work is being undertaken, and whose experiences and knowledge provide a central component of the project. Key has been separate male and female farmer cross-site visits for promoting exchange of information by gender.

Linking research to development is central to the project. A range of activities in the different countries ensures that the project benefits national conservation programmes, partner institutions and the participating farmers. The information collected is used to mainstream the use of local crop genetic resources into the agricultural development arena. In contrast to research and analysis methodologies that have widespread application, actual interventions to include management of crop genetic diversity in agricultural development activities are found to be site-specific. The project is creating a portfolio of options, based on many case studies from participating countries, to which national programmes may refer for ideas for increasing the benefits to farmers from local crop diversity.

Reflections

The method has some disadvantages. Time is needed to formulate proper hypotheses to avoid unnecessary data collection that can result because of the complexity of research and number of issues to be addressed. Integrating quantitative and qualitative data (empirical versus farmer knowledge) is a challenge. Time is also needed to build linkages between disciplines and formal and informal sectors and these linkages are vulnerable to continuing cultural, economic and environmental change.

Advantages of the research are that it focuses on locally important crops and builds on local knowledge. The gender disaggregated approach to data collection, cross-site visits and actively pursuing gender equity in employing local project team members and managers has enabled the project to more clearly address issues of livelihood improvement and empowerment and address gender and equity concerns. Farmers take ownership for their own resources. The method recognizes the achievements of both male and female farmer-breeders as 'keepers of diversity'. They in turn guarantee sustainability in that they perpetuate the process, continuing it when 'intervention' has finished. The process also channels farmers' voices into national agricultural research and extension systems and fosters cooperation between local, national and international levels and GO and NGO sectors.

Innovative approaches were developed for participatory NRM (PNRM), including developing frameworks to support the recognition, conservation and improvement of farmer-developed PGRs *in situ*. An understanding was gained of the appropriate scale for data collection, aggregation and analysis and for different stakeholders' management decisions. The process revealed the importance of taking time to strengthen farmer informal and formal linkages, integrating on-farm conservation into national PGR programmes as part of their regular annual plans and including agricultural extension staff at national and local levels in participatory training and project implementation.

Plant genetic resources are a natural resource fundamental to agricultural production. Most conservation efforts to date have focused on *ex situ* options. However, this method of conservation recognizes crop germplasm as the evolutionary product of the continuing interaction between farmers and their environments. As farmers continue planting, harvesting, selecting and storing seed, PGRs are renewed and developed. Non-use of PGRs leads to their loss from their surrounding environments. Acknowledging the dependence of crop PGRs on human use highlights the necessity of a participatory approach as inherent in the goal of conservation.

NATURAL RESOURCE MANAGEMENT RESEARCH

Since 1994, ICARDA scientists have worked with colleagues in Egyptian research institutions to design and implement a programme of resource management research in key agricultural environments in the country. Researchers carried out literature reviews, rapid appraisals, formal farm surveys and planning before establishing long-term trials at four irrigated sites (one each in the Delta and Middle Egypt, and two in the newly reclaimed desert lands, known as New Lands), and at one rainfed site (near Rafah, North Sinai). Each site has three major research problems. Water (both quality and quantity) is the paramount concern at all sites. Maintaining soil fertility is essential in the old lands of the Delta and Middle Egypt but building up soil fertility is essential for sustained production in the New Lands and rainfed areas. A third issue that the trials address is the choice of sustainable crop sequences for rotational systems. These on-station trials are designed for a minimum of 12 years and are entirely managed by researchers.

APPROACH

At each site, the long-term trial is integrated with participatory research in surrounding villages and on individual farms. Like the on-station trials, the participatory work, called long-term monitoring (LTM), is intended to have an extended life. Its purpose is to establish a continuing dialogue with farmers concerning their farming practices, management decisions and the related conditions of their natural resource base. The dialogue centres on long- and short-term farmer objectives, their perceptions of the qualitative aspects of the resource base and their technical knowledge of resource management. The participatory research also involves a longitudinal study of farmers' management of natural resources in response to changing environmental, economic and social circumstances.

As part of their exchange with farmers, researchers are also monitoring changes in the status of natural resources on representative farms through periodic biophysical measurements. Researchers are combining farmer participation with biophysical measurements to provide information about the interaction between natural resource conditions and farmers' management practices. Once institutionalized, the LTM system will provide a mechanism by which researchers and farmers can exchange knowledge on improved management practices and their effects on natural resource health. A multidisciplinary research team is conducting the monitoring at each location. Each team includes members of local farmer associations, local extension staff, researchers from various institutes and participating farmers.

These farmers were selected according to a carefully prepared list of environmental criteria relevant to each location, including hydrological and soil factors and cropping patterns. Socioeconomic factors such as farm size and type, natural resource endowment, social background, level of education and household composition were given equal weight. Farmers were selected at random from lists prepared for each site.

They received a thorough explanation of the purpose and activities of the LTM system and were asked if they would like to participate. They were also informed about the amount of time and information required and the need for a long-term commitment. The 85 farmers who agreed to participate in developing the system represent the whole range of social, economic and natural resource conditions at each study location. Without altering the integrity of the research design, the programme made provisions for new participants to join. For each participating farmer, information on socioeconomic factors, farm management decisions and perceptions of resource conditions and productivity performance are being collected every six months, after the main winter and summer cropping seasons. Natural resource conditions are measured on different schedules according to scientific requirements. In addition to basic information about crop sequences and rotations, management practices, input use, productivity and economic returns, data are collected on labour use and sources, household composition, income sources and household investment patterns. This information will explain why farmers make the decisions they do and thus should help develop profitable and sustainable production practices.

A review workshop is held once a year to bring together the research teams, including farmer members, for discussion of results and trends in the information collected.

Reflections

Through this work, Egyptian farmers, researchers and extension workers are building and testing a new holistic approach to studying agricultural production, including socioeconomic and biophysical factors and their effects on the natural resource base over time.

STRATEGIC USE OF MULTIPURPOSE FORAGES

The objective of working with rural communities is mainly to enhance interaction with local stakeholders and to achieve an impact in the initial intervention sites. The partnership with nationally based institutions is the basis for a wider distribution of results and for multiplication of farmer-selected forage seeds– seed availability being a major bottleneck for the adoption of improved forages. Conceptually, this activity relies on farmer–researcher–institutional linkages. The aims are to use forages to:

- Increase the income of smallholders mainly through direct effects on livestock production.
- Increase income and food security through, among others, improved soil fertility, soil conservation, breaking pest and disease cycles and reducing the competition of weeds.
- Ensure conservation of natural resources.

In 1998 the study began in Honduras in the CIAT reference site of Yorito, department of Yoro. Work has since been extended to other countries in Central America.

A team of researchers drawn from forage genetic resources, geographic information systems (GIS), participatory research and soils carries out the project. Together with farmers, local grassroots organizations, NGOs and other relevant institutions in and outside the region, the project develops forage-based technologies and identifies forage germplasm adapted to the needs of different types of farmers.

APPROACH

Farmers are offered a range of grass and legume options for livestock feeding, soil fertility improvement and soil conservation. For those forages immediately attractive to farmers, seed is provided in a limited amount to enhance testing, utilization and, eventually, adoption. At the same time, smallholders are encouraged to produce seed themselves. If seed production is not feasible on-site, alternative sites for production are sought.

The described approach cannot be implemented by using a single research method. The research builds on an interaction between several partners and includes the application of diagnostic tools (including both biophysical and socioeconomic aspects) to understand the production systems. To allow farmer selection of the different forage options, a combination of open evaluations and preference ranking is used.

To initiate this process of participatory selection of improved forage options, several training and planning activities with the different partners were carried out in 2000 and are continued in 2001. Several types of trials for the agronomic evaluation and simultaneous participatory selection of improved forages were established in Honduras, Nicaragua and Costa Rica. Multipurpose germplasm including grasses, herbaceous legumes, shrub legumes and cover and green manure legumes have been offered to farmers.

Thirty-nine trials were established in three reference sites in Honduras, Nicaragua and Costa Rica. The sites were selected using participatory approaches and in some cases complemented by interested farmers themselves. Each of the ten trial sites is represented by a group of 10–15 farmers. More than 100 farmers are so far involved in the project. To support this work methodologically, a procedure for the participatory selection of forages is being developed. A participatory diagnosis was done in three communities in the Department of Yoro, Honduras. Similar diagnoses are in process in Nicaragua and Costa Rica. These define the participating communities and help incorporate them in the planning process.

Reflections

Although initial results are highly promising, data have to be interpreted with care.

Results obtained in specific locations will be extrapolated using a GIS-based decision-support tool, which will be made available to a range of research and development institutions.

Preliminary results indicate that farmers selected in the first place 'improved' grasses for their pastures. Several farmers were also interested in legume species. High interest is shown in incorporating forages for soil conservation, while demand for cut-and-carry legumes is still low. However, the use of shrub legumes for dry season feeding is gaining the interest of some farmers, confirming similar experiences of CIAT research in the region. A more reserved attitude of farmers towards the legume options can be expected because in most cases the use of legumes means a change of the existing production systems.

The combination of different forage options of varying complexity is believed to be instrumental in building trust among farmers to test more complex and thus higher risk alternatives.

The community-based NRM research is seen as complementary to on-station research activities and vice versa. This study is building on this concept: the positive aspects of community-based research counterbalance the negative aspects of on-station research. For example, in on-farm research the lower control of the experimental conditions is balanced by the better insight obtained in the real-life situation.

The biggest limitation of the approach at the moment is to match the demand and supply of seed. In order to maintain farmers' confidence it is important to stress that new forage options bear certain risks and that their introduction and succesful application is a long-term process. The development of appropriate approaches is indispensable.

The approach goes beyond offering farmers a range of forage options for selection; it initiates rather a dynamic, continued process. Farmer-selection of particular grass and legume ecotypes will allow researchers to better define plant characteristics requested by farmers. Adoption of new technologies is in the first place determined by the client's requirements. These demands should help to define and direct further forage germplasm development. The interactive work with farmers may open possibilities to develop forage technology options which go beyond the immediate scope of farmers and researchers. These technologies could contribute significantly to the development of profitable and sustainable production systems. Of particular attention in this study is the search for forage options to directly or indirectly improve the well-being of less privileged groups in the rural society, such as women and small farmers (even those without livestock).

AGROBIODIVERSITY MANAGEMENT

Diversity in the natural ecological systems of mountains has contributed extensively to maintaining biological diversity in the farming systems of the eastern Himalayan region. Typically, many crop and animal species, varieties and breeds are found on farms. Subsistence farmers of the region, besides producing many crops (mainly landraces), rely extensively on wild plants to meet their needs for fibre, shelter, food, medicines, tools and household implements.

For the various ethnic groups residing in the region, survival requires extensive use and management of natural resources. More particularly, it necessitates incorporating natural resources into farming systems. By incorporating biological resources at the genetic, species and agroecosystem levels, mountain communities possess extensive knowledge of their environment.

The project is situated in the eastern Himalayas, and covers four sites: east Nepal, Sikkim, Bhutan and Nagaland in north east India. The overall purpose of the project is focused on building the capabilities of indigenous mountain populations to better represent themselves in the development dialogue. The process for achieving these objectives is to:

- Develop, through research, a better understanding of the linkages that exist between the way that gender and ethnicity are constructed and then to examine how such constructions affect the management of mountain agrobiodiversity.
- Build upon the existing experience of a network of researchers, who themselves are members of ethnic mountain communities, and increase their capabilities through skill development in research and analysis and community development concepts and practice.
- Include indigenous knowledge on agrobiodiversity management (particularly that of women) in the policy planning of national governments.

An advisory group consisting of scientists and planners from the region has been assembled to advise on effective strategies through which these can be achieved.

APPROACH

The strategies for capacity building work through research, skill development for network members and advocacy. The aims are to:

- Identify and document the extent of women's knowledge related to plant genetic resources.
- Examine changes in women's knowledge in terms of how changes in the last 200 years have transformed production strategies while introducing new crops and technologies. How have such changes been consistent with changes in ideology and how have they compromised the role of mountain women?
- Identify strategies adopted by women to counteract their increasing marginality, especially in the context of their roles as managers of agrobiodiversity on one hand and the 'gender blindness' inherent in extension and development policies of national governments on the other.

Skills of network members are developed through training and learning workshops that build conceptual clarity and analytical skills in research and help develop writing skills. Training is given on gender analysis and for facilitation and community development skills. Participatory rural appraisal (PRA) skills are included for a programme especially developed for rural planning called Village Initiated Planning. The process requires community involvement in identifying problems and in research and analysis. It includes problem prioritization, project planning (proposed action, objective, developing M&E criteria, financial management) and designing implementation procedures (methods and approaches, group formation, types of training required, guiding principles, implementation and M&E).

The advisory group, comprised of scientists, planners and development professionals in the region, propose strategies that focus on bringing an awareness of gender issues and farmers' rights to the nation's policy-makers and high-level officials via links to biodiversity planning bodies at national level and other means. The development of a participatory video documenting the process of a participatory crop improvement (PCI) initiative that builds on farmer-led approaches in eastern Nepal also helps to diffuse the method.

Reflections

While capacity building is part of an on-going process, a preliminary assessment, particularly in the PCI initiative conducted at one site, demonstrates that certain activities are more easily adopted than are others. For instance, a participatory seed management initiative conducted in eastern Nepal demonstrates the potential for dissemination of methodologies to other sites in the eastern Himalayas. The action research project was based on the following hypotheses:

- The development and enhancement of seed management technologies will occur most effectively through a process of interactive learning between indigenous and formal systems of agricultural development.
- Access to improved technologies can be most effectively sustained through community action. Thus existing technical skills for seed improvement and the organizational capacities of CBOs need to be enhanced to ensure community access to improved technologies.
- The success of community action to manage development processes will depend fundamentally on the ability to control processes of knowledge production, design and implementation of interventions.

In the 18 months since the project's implementation, members of the local CBO effectively adopted the knowledge of seed management for maize, and the community developed a seed bank with their own production of maize seed. Another significant outcome is the dissemination of the seed technology through the activities of the CBO that is presently working with over 50 farmers from adjoining communities. Lessons generated from the farmer-led seed management initiative in east Nepal have helped in developing a similar management initiative in Sikkim. Over the last few years, the spread of disease in ginger has severely hampered production of this important cash crop. The new participatory approach is based on experimenting with farmer-identified 'best' practices for disease management. A central component is through the development of local organizational capacity to more effectively participate in the research process.

Finally, members of the CBO in eastern Nepal have emphasized their interest in developing plant breeding skills, especially for maize varieties (landraces) whose seed quality has deteriorated because of factors such as the introduction of high-producing modern varieties that are available through research stations.

It is still too early in the project to assess the extent to which local knowledge of mountain communities can inform national policy planning, especially in terms of agricultural development for marginal mountain communities.

FARMER PARTICIPATORY EXPERIMENTS AND PERCEPTIONS OF CROP

Farmers use their perceptions of crop losses caused by pests to decide when to spray insecticides. This can lead to overestimating the seriousness of highly visible pests or damage symptoms. In making these decisions, farmers often rely on heuristics, or rules of thumb. Developed through experience and guesswork as to possible outcomes, heuristics may have inherent faults and biases. Farmers' decisions about leaf folder infestations in rice provide a case in point. Many farmers spray to control this pest, even though it does not cause yield losses, especially when it attacks in the early crop stages. Farmers' reactions to visible damage or insect presence may be caused by faults in their heuristics.

An approach for solving this problem is to analyse how farmers make these decisions, develop a corrective heuristic, frame it as a hypothesis and motivate farmers to participate in an experiment to test it.

APPROACH

Researchers at IRRI and ViSCA initiated participatory experiments in collaboration with technicians from the local Department of Agriculture and village leaders in Leyte, the Philippines. Half-day group meetings were held in each village for 10–25 invited farmers and a facilitator. The meetings began with general discussions about rice growing and related problems. The damage and losses that they caused, methods of control and their costs and effectiveness. Researchers facilitating the meetings eventually led the discussions as to whether control was needed at all and to the benefits of not spraying.

Next, volunteers were invited to test the heuristic 'We do not need to spray insecticides in the first 30 days after transplanting'. The volunteer farmers marked out an area of about 100 square metres in their fields that would not receive any insecticides during the first days of the crop cycle. They followed their usual practices in the rest of their fields. At the end of the season, participants reported their results in a workshop and received a certificate of participation.

Farmers from both the participating village and neighbouring ones were invited to the workshop. Pre- and post-experiment surveys were conducted to monitor changes in farmers' beliefs and intentions. These variables included farmers' beliefs, intentions, spray frequencies, timing and targets, yields, inputs and other management practices.

Rice yields in about 80 per cent of the experimental plots were equal to or greater than in the main plots. The number of insecticide applications fell from three to two per season. The percentage of farmers applying insecticides in the first 30 days of crop growth fell from 70 per cent to 20 per cent.

Reflections

The benefits of such experiments are that they are usually inexpensive and easy to conduct and they facilitate farmer learning by actively 'testing' a new idea, making participants more likely to adopt successful innovations. The approach provides a mechanism for scientists to learn about farmers' decision constraints, determine research needs, use research information and 'distil' them into testable hypotheses for farmers. It provides a means of exploring changes in farmers' beliefs, behaviour and practices. It is particularly useful for introducing a 'new' idea to a community.

Participatory experiments of this kind may have some disadvantages. They can prove expensive if the process is to be conducted over a large population. The use of media to motivate farmers' participation has been successful in Vietnam and may be an alternative to face-to-face training. Farmers participating in the experiments may risk losses and require compensation. The approach aims to introduce a testable hypothesis to farmers and thus may be viewed by some as 'top-down'.

The presence of scientists may influence farmers differently. Thus scientists applying this approach will need to acquire and use facilitating skills. Peers may view this type of participatory experiment as agronomically 'weak' because some of the controls may not be easily implemented. Because the main objective of the approach is to evaluate farmers' responses to new ideas rather than agronomy, data collection of variables on belief, behaviour and practice changes needs to be emphasized.

The adoption of innovations such as seeds and machines is often discussed in the literature.

Less is known about the adoption or adaptation of information into farmer decisions. Because much of resource management is in the form of information and how to adapt and integrate it into decisions and practices, investing in decision research, an emerging field of applied social psychology, will enhance delivery and communication information.

FARMERS' ABILITY TO MANAGE A DEVASTATING PLANT DISEASE – POTATO LATE BLIGHT

Resource-poor farmers have substantial difficulty in managing the diseases that affect their crops. Potato late blight (LB) is particularly devastating for small-scale producers. Because of recent worldwide migrations of more virulent and fungicide-resistant strains of the pathogen, potato farmers face a problem that behaves differently than before. Poor farmers have little knowledge of the disease, in part because the organism that causes it is essentially invisible. Late blight is usually managed through the use of fungicides, some of which are suspected carcinogens. In developing countries, effective disease management strategies are best devised locally, because of the tremendous variation in human, environmental, host and pathogen factors among potato agroecosystems.

As the result of decades of resistance breeding, potato varieties and breeding lines with promising levels of resistance are available. Although efforts are made to breed for durable resistance, varietal diversification is desirable to reduce the erosion and breakdown of resistance. Getting improved varieties to the farmers is, however, a significant challenge because of the limitations inherent in a vegetatively propagated crop. Deployment of promising breeding lines in marginal and heterogeneous environments without formal seed systems is particularly difficult. Participatory approaches are essential because they help integrate varietal selection with other elements of disease and crop management strategies, and they contribute to the improvement of informal seed systems.

The Food and Agriculture Organization (FAO) developed the farmer field school approach (FFS) for training in IPM. Since 1997, CIP has been working with several research and extension institutions to develop and implement FFSs with farmer groups in the Andes and elsewhere. The original FFS approach was found to require substantial adaptation for potato. In potato production, farmers make many of their key decisions before the start of the growing season. The variety that they choose and their source of seed are key issues in managing LB and other potato pests. We incorporated a substantial element of farmer participatory research (FPR) into the FFS format, and therefore designated our approach 'FPR-FFS' to distinguish if from the FFSs that focus principally on training.

APPROACH

To initiate the development of an FPR-FFS focused initially on potato late blight, CIP convened a series of local and national meetings and an international workshop to develop a strategy and to define available materials. An FPR-FFS curriculum, embodied as a field guide for facilitators, was drafted. A baseline study on LB was conducted in Ecuador, Peru, Bolivia and Uganda. This study confirmed that LB is the most important production problem for potato farmers and provided insight into farmers' knowledge and practices.

In an FPR-FFS, a group of about 10–35 farmers from a given locality meets regularly over the course of one to three cropping seasons (or longer), usually twice a month for a half-day session. Helped by a trained facilitator (usually from a local NGO), the farmer group conducts field experiments and hands-on learning activities. They use direct experimentation and observation to improve their knowledge, and use this expertise to improve their crop and pest management. The field experiments have included testing of promising varieties and/or breeding lines, testing different fungicide strategies for varieties with different levels of resistance, working with varieties derived from true potato seed and comparing IPM and conventional practice. By sharing data among communities through field days and workshops, the groups amass substantial data and can proceed with decisions such as varietal selection with relative confidence.

The programme has expanded in Peru and elsewhere. In 1997, CIP and CARE-Peru initiated FFSs on a pilot scale in four communities of San Miguel, Cajamarca, in northern Peru. Eight FPR-FFSs were conducted during 1998–1999 in San Miguel, and 13 were conducted in 1999–2000. In parallel, pilot-scale FPR-FFSs were established elsewhere in Peru, as well as in Ecuador, Bolivia, China, Bangladesh, Uganda and Ethiopia through the collaboration of researchers, extension organizations (mainly NGOs) and farmer groups. Parallel efforts by other organizations in coordination with CIP led to further efforts in potato. Facilitators from Bolivia, Ecuador and Peru were formally trained in FFS methods through the FAO, and national IPM projects emphasizing the FFS methodology were established in Peru and Ecuador.

Reflections

Farmers have much to learn about the microbial world. Because they cannot see the organism that causes plant disease, they do not understand disease processes well. They are poor at diagnosis and inefficient at managing the diseases that affect their potato crops. However, given the opportunity, they are quick to learn and improve their management decisions. They are keen to try new varieties and are appropriately conservative about making decisions regarding varietal change. Participatory evaluation gives farmers a meaningful basis on which to make decisions about varietal choice.

The FPR-FFS approach is demanding on both farmers and researchers. Farmers must be strongly motivated to improve their potato production if they are to participate successfully. Because potato is a high-value crop and the losses caused by LB are often devastating, LB is a suitable entry point. However, farmers face numerous problems with their potato crops and other agricultural enterprises, and prefer integrated approaches that allow them to cope with multiple problems at a given time. The FPR-FFS method forces researchers, who often have narrow technical interests, to expand their horizons. Researchers and farmers have complementary roles in the evaluation of potential new varieties. Linkages between research and extension organizations increase the potential impact of knowledge-intensive technology.

In the Andes, the FPR-FFS has attracted many young men while older men and women with young children often found participation more challenging. This could result, in part, from the limited role that women play in potato production in the northern Peruvian Andes. However, women have a stake in successful crop and disease management for potato, and in selection of appropriate potato genotypes.

In participatory evaluation of varieties and breeding lines conducted through the FPR-FFS, the opinions of men and women participants were sometimes significantly different.

This reinforced the importance of involving both men and women in the activity. Efforts are being made to improve the training curriculum to enhance its utility for female participants.

The FPR-FFS approach is still evolving, and from the outset has had much in common with the CIAL methodology. With support from the International Fund for Agricultural Development (IFAD) and the Organization of Petroleum Exporting Countries (OPEC) Fund for International Development, pilot-scale FPR-FFSs are now being established in seven countries, through collaboration among researchers, NGOs and farmer groups. More emphasis is being placed on gender analysis (assisted by the PRGA programme), and the impact of the FFS is being assessed (PRGA and the World Bank). Preliminary observations indicate that the FFS is highly effective in stimulating farmer learning and varietal diffusion.

FARMERS' RISK MANAGEMENT STRATEGIES

Small farms of less than 5 hectares account for about 70 per cent of southern Africa's maize production. Although new technologies are available for improving production in smallholder maize systems, widespread adoption faces major constraints – particularly the constant threat of drought and declining soil fertility. In Zimbabwe and Malawi, the soils in smallholder areas tend to be sandy, with limited organic matter, low nutrient content and low water-holding capacity. Farmers have only limited access to organic manure and cannot usually afford inorganic fertilizers.

The threat of drought, combined with fluctuating market prices, means that farmers are gambling on an uncertain yield and economic return. Therefore, to be attractive to farmers, new technologies for improving soil fertility must be able to reduce production risk. They must also be compatible with farmers' diverse and complex livelihood strategies. To support the development of appropriate soil fertility technologies, CIMMYT's Risk Management Project (RMP) evaluates their biophysical and socioeconomic performance through a combination of computer crop modelling and farmer participatory research in Malawi and Zimbabwe.

APPROACH

The RMP employs both hard (quantitative) and soft (qualitative) approaches to explore the links between agroecosystems and the socioeconomic environment. A participatory research subproject conducts systems diagnostics, identifies stakeholders, determines farmers' soil and climate taxonomies, describes farm families' livelihood strategies and fosters farmer experimentation with soil fertility management practices. A modelling subproject collects data to validate the computer model and fosters its use to examine the long-term biophysical performance of soil fertility management practices under specific soil and climate conditions.

By integrating the two sets of activities, RMP can use farmers' soil and climate taxonomies to develop soil and climate profiles for running the model. Further, the model can be used to evaluate farmer-developed technologies, and farmers can evaluate the model's outputs and participatory technology field trials within the context of their livelihoods and risk management strategies.

This enables farmers with diverse resource endowments and diverse livelihood strategies to verify options within their own environments.

The RMP collaborates with focus groups from the Universities of Zimbabwe and Malawi, national agricultural research programmes and the Africa Centre for Fertiliser Development. Researchers and farmers evaluate soil fertility technologies being developed by the focus groups. Through this integrated approach, researchers can draw on one another's experience and that of farmers. The RMP also has links with ICRISAT and CARE. The CIMMYT and ICRISAT centres jointly fund researchers and field activities and share information. Farmer groups established by CARE link the RMP to communities and provide a social framework for broader dissemination of successful technologies.

One of the activities is to design a framework for running simulation models based on farmers' soil management practices. The goal is to develop an interface that permits discussion of outputs and key management variables, involves farmers in assessing scenarios developed by the model and enables them to ask questions of the model.

Focused planning meetings are conducted to enhance the team's organization. Participants develop common work plans and research frameworks for all project stakeholders. The RMP began with a macro-systems diagnostics approach, which enabled it to identify key stakeholders, secondary data and partners for implementing project activities and to identify appropriate techniques for fieldwork. The RMP has thus created a strong network of research partners throughout the region.

Fieldwork activities concentrated on forming or strengthening farmer groups at two sites in Zimbabwe and one in Malawi. Farmers, extension staff and researchers formed new groups for the 1999–2000 crop season. Activities included participatory wealth ranking, the development of farmer taxonomies of soils and climate, inventories of management options and practices for different resource endowments and varying soil and climatic conditions.

A key collective learning and decision support tool is the use of participatory modelling maps of agroecosystem resource flow or resource allocation maps (RAMs).

These demonstrate key household and community resources and inter-relationships within a systems context. Farmers and researchers together develop the maps and use them to record, monitor and analyse data and decision-making, which then enables them to understand the systems' context of soil fertility. The methodology development for the modelling and FPR interface is considered an iterative and dynamic process, with a diversity of tools and techniques being used, refined, adapted or discarded as the process and project stakeholders require. This is an immensely creative and ambitious research agenda. This kind of interaction is highly unusual but holds great promise for a more effective evaluation of soil fertility management technologies, under highly variable and risky climate conditions.

Reflections

Biophysical crop modelling enables evaluation over time, including different variables such as climate, soils and management. However, the models are only as good as the data and the agenda that are used in their development and are highly data-intensive to establish. Models are often narrowly focused compared to farmers' multi-faceted and complex livelihoods and cannot handle socioeconomic variables. Some factors affecting crop and soil performance are also beyond the model's capabilities. The model-to-farmer interface integrates disciplines and can bring modellers and farmers together with quick feedback.

This can be used to promote collaborative learning and as a decision support tool because it makes it easier to identify and target key research priorities. It has some knowledge gaps with outputs highly subject to interpretation as to by whom and how the research agenda is driven. Linking modelling with GIS and farmer land types for scaling up is being explored but entails expensive start-up costs and questions of data ownership.

On-farm participatory testing of technologies allows farmers more freedom to experiment and helps bring stakeholders together. It serves to develop a better understanding of farmers' priorities and natural resource and socioeconomic factors affecting technology performance, also capitalizing on farmers' indigenous knowledge. On-station testing is easier to manage and provides rigorous controls and designs that reduce variables affecting crop performance. However, socioeconomic variables associated with management are excluded and a long time frame is needed. This integrated research method and process enables a learning forum to be developed whereby the very different mindsets and approaches of all the actors involved in the research can begin to envision a common environment. They can then develop a platform to work together on common problems and solutions within a participatory process

FARMER PARTICIPATORY RESEARCH METHODS

In Malawi and Zimbabwe, ICRISAT is developing new research methods for improving soil fertility. The Institute is also building partnerships among national scientists, extension advisors in NGOs and the public sector and farmers. The aim is to improve the ability of national research programmes to develop 'best bet' NRM technologies for poor farmers by obtaining their input at earlier stages. New technologies are needed that improve human nutrition while enhancing soil management and enabling communities to rehabilitate degraded environments. The main innovations introduced so far are intensification with long-duration and indeterminate legumes and integrated use of organics and inorganics. In Malawi, farmers are testing and adapting options such as maize grown in rotation with a doubled up legume intercrop of pigeonpea and another grain legume. Combining small amounts of fertilizer with manure and pigeonpea residues is also being tested.

APPROACH

A novel aspect of the programme is its evaluation of several participatory approaches applied in parallel in different villages. The results are compared with baseline data from villages having no known relationship with researchers or farm advisors from NGOs or extension services. This approach enables researchers and farm advisors to address their concern that farmer adoption of fertilizer and integrated nutrient management has been practically nil, despite a decade of on-farm research and the recent focus on participatory research and extension as well as training-for-transformation empowerment approaches.

Costs, as well as benefits, will be assessed for different methods of partnering. Some researchers and senior extension staff are also concerned that extension rarely reaches female-headed households and women farmers, nor do their concerns enter into agronomic research.

The methods being compared include farmer empowerment approaches led by NGOs, extension-led demonstration trials and farmer participatory research, each conducted in a different village in the case study areas. The study is determining which of the approaches are best at building institutional linkages and improving the peer relationship among stakeholders, for different locations. All of the partners involved evaluate the effectiveness and costs of each approach. Researchers are also determining how well each approach addresses the needs of female-headed households. Project partners conducted comprehensive surveys to provide a baseline for, and developed methods of, comparison. They agree that the comparison should indicate which methods are working best, as reflected in perceived cost-effectiveness and the satisfaction of researchers, extension advisors and farmers. Farmer adoption and adaptation of technologies, farmer empowerment and improved soil management will be assessed for each method.

The 'mother–baby' trial design is one method for improving communication between farmers and researchers that has so far proved successful.

The approach was originally created to facilitate farmer collaboration in testing soil fertility technologies. Further, it was conceived as a practical means for researchers to rigorously incorporate farmer evaluations of technologies at every step in the development process. The approach links together two trial types: a replicated one that fits researchers' needs and a simpler trial that meets farmer needs, rather than attempting to compromise and meet all objectives in a single trial.

Researchers first design 'best-bet' technologies, attempting to take into account farmers' priorities and resources. Then mother–baby trials are planted in each participating village. The 'mother' is a replicated experiment designed by a researcher. Farmers plant and manage the baby trials, which are single replicates of the mother trial. For this purpose, each farmer selects a best-bet technology from the mother trial and adjusts the level of inputs and equipment according to his or her preferences. Farmer evaluation is documented through surveys, community discussion and ranking exercises, which facilitate researcher incorporation of farmer input. The trials have led to improved spontaneous experimentation among farmers and give researchers and extension advisors the chance to observe and learn from farmers.

Reflections

The larger context of the different case study areas where the approaches are being tested is difficult to assess. For example, the market opportunities and historical extension–farmer relationships may vary markedly among the areas and determine the relative success of different methods. The trial and demonstration approach probably has the least emphasis on farmer knowledge, while facilitation of farmer learning and experimentation, the farmer-led approach, has the most. The changes in the researchers and extension staff to encompass a broader, more participatory approach is proving difficult to document. Further steps, such as how to facilitate farmer training and communication with other farmers on guidelines, are not explicitly part of any approach although this may develop. Farmer-to-farmer communication is probably the most important means of technology dissemination and is not treated explicitly in any of the approaches, except perhaps the farmer empowerment approach (which some professionals view as too costly for large-scale work). This point is difficult to evaluate but should become clearer with time. A survey is underway to evaluate researcher and extension attitudes and beliefs regarding effective ways to communicate with farmers and work together to develop soil management options and to improve farmer experimentation. Surveys documenting farmer assessment of the process and farmer adoption are also on-going.

All of the partners involved in the methods' comparison systematically evaluate methodology approaches and efforts to facilitate farmer–researcher-extension linkages and technology best-bet options. Also, the concerns of women farmers and female-headed households are specifically addressed. This took time to build, in part because almost all of the researchers and extension staff involved were men and no gender sensitivity training or discussion was attempted at initial stakeholder meetings.

Attention to including women farmers in the technology development process was almost nil at first but has increased over time, particularly at the Malawi sites. This adds value to on-going efforts in the area. The mother–baby trial and farmer empowerment approaches attempt to facilitate farmers' learning basic research principles, to expose farmers to a range of new options and to empower them to value their own knowledge. This appears to have improved communication among farmers, researchers and extension staff.

A conundrum is that extension staff and researchers mostly see the trial demonstration approach as the only cost-effective way to scale up dissemination.

Yet this approach does not facilitate farmer experimentation or joint learning. Staff at NGOs also state that their work on farmer and community empowerment is only cost-effective in isolated areas. They believe that they need to go to trials and demonstrations to reach more people, although both groups express frustration with the lack of effectiveness of demonstrations led by extension or farm advisors.

The goal of this case study is to measure long-term impact and changes in how research and extension staff conduct their work. Stakeholders designed this village-based comparison of methods and are involved in evaluating the pros and cons of each participatory method over time. Meetings are held annually for assessment. Strong links are built because researchers, extension staff and NGOs were already carrying out most of this work and through the case study we are attempting to facilitate self-reflection on the value of different approaches to the stakeholders. In Malawi, 400 farmers are assessing best-bet technologies at seven sites around the country through baby trials and their own experimentation. In the process, they are satisfying researchers' need for sound quantitative experiment results.

COLLABORATIVE MANAGEMENT OF NATURAL RESOURCE MANAGEMENT

Tsetse control operations commenced in the Chiawa area of Zambia in 1991. Prior to that date, no cattle were kept in the area because of tsetse, although cattle were kept in surrounding tsetse-free areas. Tsetse control dramatically increased the agricultural options in the area but also presented the challenge of how to realize the benefits without over-exploiting the environment.

Typically, 'land use planning' is seen as the means of meeting this challenge but the record of conventional approaches to land use planning under communal tenure systems in southern Africa has not been impressive to date. Over the period 1994–1997, Chiawa was selected to pilot alternative approaches. The overall objective was to increase stakeholder involvement through decision-making support in order to achieve collaborative management of natural resources.

APPROACH

In Phase 1, community leaders and government and NGO representatives were consulted. Scheduled community meetings were held. A review was made of existing documentation.

Rapid air-photo interpretation and rapid field surveys of farming systems, vegetation and soils were also carried out. The results of the field surveys were incorporated into a simple GIS. Consultations revealed that an alternative approach was needed with improved stakeholder involvement as the primary objective. They clearly identified the importance of responding rapidly to demand-driven initiatives as opposed to imposing externally conceived 'solutions'. Thus, a 'demand-led and support' model was developed in contrast with the more familiar 'suitability and enforcement' model of land use planning.

The approach was based on the following principles:

- The community, not outsiders, proposes the planning initiatives.
- The community evaluates initiatives and assesses them for environmental impact.
- Coordinators from *within* the community run the scheme.
- The community undertakes most of the implementation. External support is largely confined to technical advice and loans of equipment.
- Communities must demonstrate 'commitment' to projects before any external support is provided.

Female and male 'local community workers' (group promoters) were recruited by advertising locally and selected by interview. Training was provided that included periods of attachment to experienced community workers within Zambia. By clearly explaining the principles of the scheme to all involved from the outset, false expectations were avoided and 'ownership' of the initiatives reinforced. In Phase 2, the community workers convened a series of small meetings at which problems and solutions were discussed.

A scheme for supporting small projects that were both maintained by the community and environmentally benign was jointly developed. The process of local approval was agreed during these initial meetings. A wide range of projects was proposed, including fishponds, cattle paddocks, vegetable gardens, handicraft production, marketing of local tourist attractions, well construction and poultry production schemes.

The proposals were documented and passed to the relevant authorities for evaluation. Environmental impact was assessed during site visits and from information collected during Phase 1. Approved projects were supported through a variety of means, including technical assistance, loans of equipment and short study tours.

Interaction with the community during this process provided valuable guidance for parallel components of the planning process. Phase 3 addressed issues of communal grazing resources, boundaries and procedures for arbitration of disputes. Trust, understanding and confidence were built during the scheme's development that enabled more productive debate and subsequent actions than would otherwise have been the case. Phase 4 examined options and outlined costs for community infrastructure provision such as roads, schools and clinics.

Reflections

Three main lessons were learned:

1 The process takes time. Frequently, land use planning consists of a time-bound period of data collection and consultation. This can result in distortions in the perceptions of planners. The presence of community workers over a longer time scale serves to correct rushed generalizations. Their involvement and continuous presence serves to maintain momentum and act as a two-way conduit between the community and others.

2 The experience and capability of the local coordinators has a key influence on the direction of the planning effort. Careful selection, support and training are essential. The disadvantages of selection from the local community include lack of objectivity, favouritism and possibly diminished respect because of familiarity. The advantages of local selection overall outweigh the disadvantages and include a positive slant on these in terms of understanding and familiarity combined with the commitment to improve their own community.

3 Initially groups, as opposed to individuals or families, were targeted because of the greater numerical impact. Typically, group initiatives started well and initially made rapid progress but, in several cases,

social tensions reduced their effectiveness over time. In contrast, individual and family initiatives took longer to start up but once established were all sustained.

A 'demand-led and support' model was seen as an effective entry point in developing a participatory land use planning process, initiated by stakeholder involvement. The approach provided decision-making support and encouraged the collaborative management of natural resources. The community enthusiastically received the scheme. Eighteen months after its initiation, 35 projects were operating independently. The transfer of coordination to a local NGO demonstrated the success of the approach. Proposals to establish a small revolving fund to support further initiatives were made. The individual projects were likened to pieces of a mosaic. The projects have individual integrity and ownership, but when placed together they constitute a de facto land use plan.

PARTICIPATORY MANAGEMENT IN LANDSCAPE

From the top of the hill we could see the landscape far into the hazy horizon. Dark clouds threatened a thunderstorm, so we quickly finished drawing our map of the territory and moved down the immense granite stone where we had been standing. We had been discussing problems of the local wetland – not easily visible from our vantage point because it had shrunk so much to make room for rice fields. Stanley mentioned that since my last visit, seven years before, people had introduced several innovations. They reduced tree cutting and planted trees in the boundaries between their fields. Many families had successfully tried out ways of intensifying agricultural production. They had built contour ridges around their fields, introduced black ants that feed on aphids, spread ashes on crops, weeded at better times, selected their best seeds for planting, and introduced new activities such as rabbit raising or bee-keeping. Some had even gone so far as to plant the highly beneficial but difficult-to-raise *neem* trees. Yet, the government was still encouraging rice growing by individuals, and the wetland had kept shrinking, seemingly on its way to total disappearance.

Amos began explaining why this was a problem. 'When the big rains come, the wetland acts like a sponge. Before, even in heavy rains, it protected our village from flooding. Now the wetland has shrunk so much that even a little rain floods the fields and our homes in Kapuwai.' During the dry season, water shortage is becoming more critical. Daniel said, 'We need more watering points for people and animals. We need more fish from the wetland and the medicinal plants that grow close to it and are now getting hard to find.' Extensive rice growing was causing problems of all kinds, from loss of biodiversity to the loss of schooling for the children who spent days surveying the fields for birds.

The community was convinced that the wetland was important for everyone and needed to be used carefully and not destroyed. However, members were still unclear about what to do. Anne said, 'our Association dreams of managing the wetland together, as a community'. People felt that they could do so by taking advantage of their traditional management skills and of the skills the local Association had acquired while managing common resources in agricultural production.

That evening, we ate together in the light of some candles and paraffin lamps that sent a pungent smell into the small room of the health centre. Later, more people joined our small group and we held a larger meeting. The topic of discussion was the management of the wetland. We went far into the night discussing the beginning of a positive vision that the whole community might be willing to share.

APPROACH

We met again the next morning. The participants in the evening discussion agreed to call a larger meeting where the whole community would discuss the vision for the future of the wetland. In particular, they agreed on calling various 'stakeholders' to present their individual views and to negotiate a basic agreement on rules to be respected and activities to be carried out. We identified three key elements in the participatory management process for the Kapuwai wetland:

- preparing for the partnership;
- developing the agreement(s); and
- implementing and reviewing the agreement(s)/'learning by doing'.

In the first phase, discussion on the wetland issues ('social communication') would be promoted in the community. In this preparation time, people would begin to clarify what they needed to know for sound management and the reasons why the wetland was useful for them. In the second phase, stakeholders would be invited to set aside their immediate interests and develop together a long-term vision for their wetland. A facilitator would help at this stage. Once a common vision was reached, a ritual would make it intangible and sacrosanct. Clan elders and traditional authorities would perform the ritual and every 'stakeholder' would be asked to re-affirm the desire to work together to reach the common vision.

After the ritualization of the common vision, it would be time to negotiate a management plan, some basic rules for the use of wetland resources and other needed accompanying initiatives. All stakeholders would be invited to meetings and they would strive to work by consensus – not by majority vote – and to be totally transparent about information and all kinds of decision-making.

In the third phase, the committee would be acting on the basis of its duties, and the agreed plans would be implemented. This would be far from supine implementation, however! The Kapuwai people stressed that implementation should be taken up as a way of 'learning by doing'; thus they would have to plan in advance for regular reviews and discussions of management results. They felt that the whole community should be allowed to participate in such meetings.

Reflections

We were conscious that several difficulties needed to be overcome for successful results. A few members of the local Association and others needed to make a substantial investment of their time to set up the process that would lead to participatory management. The interests of landowners needed to be respected but also seen beside the community interests as well. Many thought that the Council of Elders should be involved from the very beginning of their wetland initiative. Alternative ways of raising income (in place of rice cultivation) needed to be provided. Effective facilitation and support was also to be secured during the organizing and negotiation process.

We expected the participatory management of the Kapuwai wetland to increase the sustainable use of the wetland resources and to protect the biodiversity remaining in the area. A more equitable use of wetland resources was also to produce negotiated specific benefits for all the stakeholders. Early discussion and agreements on the regulations of wetland use would prevent social conflicts and problems. Other envisaged benefits were the enhancing of local capacity for wetland management, the fending off of exploitation from outside interests and the enhancing of the vitality and 'identity' of the Kapuwai community.

When the time came for us to leave, our friends' eyes were lit with enthusiasm. They had just charted for themselves a way towards a better future for their wetland. With luck, patience, and personal effort, they would succeed

in following it up together, for the benefit of everyone. I still vividly remember when they waved us goodbye with broad smiles on their faces, surrounded by a large and cheering group of family members, friends and children . . . The above account refers to July 1999, more than three years ago. Since then, I received some information via email, as Stanley manages once in a while to travel to Kampala.

Our friends – who are organized in a local association called PACODET – set up several workshops in ten relevant local parishes. The workshops were widely attended (women representatives, clan leaders, local leaders, church leaders, youth representatives, extension workers, district councillors . . . more than a hundred people in each workshop!) and dealt with what should be done with the wetland, also in the light of the national Land Act provision.

The majority of participants were not happy with the practice of clearing and draining large wetland areas for rice growing, and agreed this was killing the wetland and denying to the communities benefits such as water, grazing, fish, firewood and grass for thatching. They also reported that many water sources had dried up because of the wetland draining. The Land Act provision states that wetlands are a public resource, which should be sustainably used to the benefit of all stakeholders. The local authorities were requested to support the implementation of this provision, but some of them own land in the wetland areas and are opposed to it.

In the meantime, an Environment Officer was recruited by the district, and environment committees were called to form at all levels, from the district to the parish. The association of our friends decided to work within the newly created system and concentrated efforts to lobbying key district and sub-county officials, environment committees and the Environment Officer to support the de-privatization of the wetland and its management and sustainable use by the local communities. PACODET also identified a number of activities that could be demonstrated in and around wetland areas, in support of both livelihoods and conservation and wrote proposals to raise funds to set them up.

In a nutshell, the local association has done the groundwork of social communication on the issues and is now focusing its energy on lobbying the local authorities for supportive action and mobilizing resources for community-based demonstration activities. The main obstacle to continuing their work in participatory action research for the management of the local wetland is a clear one: private ownership of land.

PARTICIPATORY RESEARCH AT THE LANDSCAPE LEVEL: THE KUMBHAN WATER TROUGH CASE

The present case study is based on collaborative work between NRI and BAIF Development Research Foundation on the research project 'Easing seasonal fodder scarcity for small ruminants in semi-arid India, through a process of participatory research'. The project was funded by the UK Department for International Development's Livestock Production Programme.

In the project village of Kumbhan in Bhavnagar District, Gujarat, livestock-keepers said that seasonal water scarcity was their main constraint: mean annual rainfall in Bhavnagar is about 500 mm and is concentrated in the period of July–September. During the summer season the number of water points is limited, and there are none near the main grazing area, so herders and their animals are forced to walk long distances. The researchers considered water and feed scarcity constraints to be inter-related, and hence research on water scarcity was within the remit of the project.

The Rabari livestock-keepers proposed the construction of a water trough and storage tank adjacent to a privately owned well, in the vicinity of the *main dry season* grazing area. The owner of the well was agreeable to supplying water to the trough. The trough was constructed in April 1999, in time for use during the late dry season in May and June. The researchers monitored various parameters before and after the trough came into use.

Gujarat is a vegetarian state in which meat production and consumption are socially unacceptable in rural areas. Thus, milk and manure are the main livestock products. The Rabaris specialize in livestock production (mainly cattle and goats), and livestock herding is the full-time occupation of some male Rabaris This group was keenly interested in the work from the outset, because it addressed the priority livestock production problem that they identified, and because they had proposed construction of the trough.

APPROACH

The water scarcity issue was raised during a semi-structured group interview with Rabari men, as part of the initial survey work on livelihood system characterization and needs assessment. They identified and ranked their main livestock production constraints as:

- water scarcity (dry season);
- feed scarcity (dry season); and
- disease.

Frequent informal contact occurs between BAIF local staff and the villagers because BAIF has an office in Kumbhan and is involved in other development activities there. This provided additional information. Livestock production constraints – and the relationships between causes, core problem and effects – were further elucidated through a participatory problem tree analysis undertaken by Rabari men in November 1998. It is a diagrammatic tool for analysing problems and gaining a more in-depth understanding of their nature. This is important for assessing the implications of interventions: because constraints are often inter-related, easing one or more can lead to the alleviation or exacerbation of others. The tool involves identifying a core problem, the factors causing it and its effects.

The problem tree constructed by the Rabaris incorporated both biophysical and socioeconomic factors and showed how they inter-relate. The Rabaris identified reduced milk production and disease as two specific effects of water scarcity in the dry season. They also identified the impact on themselves (ie, walking considerable distances in the intense heat, with lack of drinking water at times, leading to exhaustion at the end of the day). They expected a general improvement in the performance of their animals due to the saving of energy from the reduction in herding distances. Before deciding whether to proceed with construction of the water trough, the local BAIF staff collected data that would enable an informed appraisal to be made. The data included:

- Current daily herding routes and distances and livestock-keepers' estimates of the effect of the trough on these.
- The number of herders and livestock (by type) expected to use the trough.

A BAIF consultant made a detailed estimate of the trough cost. The NRI socioeconomist used this to do a simple cost–benefit analysis, in which the benefit was expressed in terms of time saved by herders. This suggested that the trough would pay for itself in little more than one dry season.

The researchers wanted evidence of the livestock-keepers' commitment from the outset, and wanted them to be responsible for the trough in the future. Thus, the latter provided the construction labour voluntarily; and also agreed to form a management group that would take full responsibility for the future maintenance of the trough.

Evaluation meetings were held in late July – with Rabari women and men separately – at which they were asked for their views and observations on the impact of the trough.

Reflections

The evaluation meetings confirmed that the expected benefits to both animals and herders had been realized. The women revealed that before the water trough came into use their husbands tended to be tired, irritable and argumentative when they returned home in the evenings. Since then they had been less irritable, and if there was a disagreement between husband and wife it could be resolved amicably.

The use of participatory problem tree analysis proved highly useful. It reveals how farmers or livestock-keepers perceive problems and relationships, which may be different from how outsiders see them. For example, livestock scientists tend to focus on how constraints affect the animals, whereas these livestock-keepers were also concerned about the impact of water scarcity on themselves.

There were some problems with the collection and analysis of monitoring data. First, the design of the monitoring system was researcher-dominated: the Rabaris themselves did not consider it necessary to collect such detailed quantitative data. Second, the BAIF field staff were not used to conducting research and did not analyse the data themselves. As a result, they were unaware of puzzling differences in milk production trends that could have been usefully discussed with the Rabaris. This highlights the need for field staff to be proficient in simple techniques for analysing and inspecting monitoring data.

AN INNOVATIVE COMMUNITY APPROACH

This case study is set in the mid-hills of Nepal and draws on the experience of researchers, extensionists and farmers in developing and implementing an innovative community approach to the control of bacterial wilt (*P solanacearum*) of potatoes (*S tuberosum*). The architects of the approach were the development scientists of Lumle Agricultural Research Centre.

The high hills of Nepal have traditionally been a source of supply for seed potatoes to the mid-hill and lowland (*terai*) potato producers of Nepal because of the low incidence of viral diseases in the high hills. Bacterial wilt is a serious disease that can survive in the soil for several years and can be spread through infected seed potatoes. It threatened the trade in seed potatoes (and the production of potatoes as an important hill staple) from the late 1980s, as it became established in the villages where seed was produced. The villagers themselves did not know the life cycle of the disease, nor what control measures to take. Lumle Agricultural Research Centre held the research mandate for the area and devised a strategy for addressing the problem in collaboration with the affected communities.

The whole process has been an integrated effort between natural and social scientists working together with communities. However, the Samuhik Bhraman carried out at the start of the project was probably the key to understanding the interaction between the social, cultural, physical and biological factors.

APPROACH

The Samuhik Bhraman is a type of rapid rural appraisal (RRA) that evolved in Nepal during the 1980s and involves a multidisciplinary team of researchers, extensionists and villagers in exploring a defined subject. In this case, it was focused on the seed-potato production system. The appraisal uses a range of RRA/PRA tools, with team members changing each day between subgroups to promote cross learning. Each evening a reflection period is held when subgroups discuss what they have learned and decide tasks and responsibilities for the next day. The team lives in the community for the duration of the study and confirms its findings through community meetings. Through this method it was possible to define the geographic and temporal distribution of potato production, the production and storage methods used, the constraints and the economic imperatives. Relevant social and ethnic structures and cultural practices (such as exchange of potatoes as presents and the sharing of tools and livestock for draft) were also identified.

Four seed-producing villages were selected, with contrasting social characteristics and size. A Samuhik Bhraman confirmed bacterial wilt as a major problem. Major reasons for the fast spread of the disease were lack of awareness of the disease, frequent movement of potatoes between and within villages, short crop rotations, poor plant hygiene and the use of volunteer potatoes for tuber yield. A multidisciplinary team comprising phyto-bacteriologists, agronomists, extension workers and socioeconomists devised a plan for management of the disease in 1990. Farmers were involved in monitoring the disease, and meetings held with villagers to create awareness of it. Each pilot village created a 'Cropping System Improvement Committee', which was responsible for the programme within its village. The project took special pains to provide training to women and men of different social groups.

Key components of the approach to integrated management of the disease were:

- Eliminating infected planting materials from the village.
- Providing a pathogen-free seed multiplication programme in the community for a regular supply of healthy planting materials.
- Prohibiting cultivation of potatoes or other solanaceous crops for at least three years in infected fields.
- Rouging of volunteer potatoes.
- Educating farmers on the symptoms of the disease, its transmission, control measures in field and store and sanitary aspects of disease management.
- A support programme bringing alternatives to potato production.

To compensate for the loss of potato production, alternative (non-host) crops had to be provided. Demonstrations of nursery raising and vegetable production, seed supply and technical advice were therefore important components of the support programme to project villages. In addition to vegetables, cold-tolerant rice was to become an alternative to potatoes. The posting of a facilitator/extensionist to each of the project villages as liaison between researchers and the Committees assisted the process.

Reflections

Control of bacterial wilt is technically feasible. However, it was difficult to achieve the required level of community participation essential to ensure long-term success. Success in containing or eliminating the disease varied among villages before 1996. Jhilibarang village, where community cohesion was strong, continued disease-free seed potato production for the three years of the project. In Ulleri, community cooperation was difficult to manage, and the disease appeared from year 2 of programme implementation. The programme was terminated in Ghandruk after year 2 (a village with a lucrative tourist trade and less dependent on agriculture) and the disease reappeared in Sabet when farmers resumed their normal cropping patterns and grew potatoes in traditional fields. The communities vary greatly in their levels of cohesion (those with greater 'jaat' (caste) diversity were less cohesive) and the socioeconomic environment and the existence of alternatives to agriculture (eg, incomes from tourist trekking routes) reduce the need for community compliance. The high dependency on potatoes makes some farmers take action for short-term gain that results in long-term disaster. Constant changes in social equilibrium and the influences of exogenous and endogenous forces require careful monitoring and response.

Village workshops, training and cross visits need to be provided to broaden the level of thinking and to improve participation based on understanding. These have to be sensitive to the needs of women and the poor who cannot easily leave their duties to attend training. It has to be recognized that 100 per cent cooperation is very difficult and grievances of non-cooperating members need to be targeted and understood. Key elements to success are a coordination mechanism, a monitoring system and a supportive policy framework. Scaling up from the pilot project to wider application requires comprehensive information followed by a massive awareness/training/support programme.

The method leads to community cohesion and an effective use of awareness raising and training. Good support is given by a well-established and well-respected research and extension service. A further benefit is the identification and support of alternative NRM options with excellent multidisciplinary support from technical and social scientists.

INNOVATION IN IRRIGATION MANAGEMENT

Irrigation management has some particular challenges in participatory research because it involves collective action where different interests may be present. Large-scale irrigation systems create particular challenges in participatory approaches to innovation, because of the large number of water users and the different ideas on participation and innovation present with different stakeholders involved in transforming irrigation practices. Using examples from an irrigation management reform programme in the Terai of Nepal, this case study summarizes operational elements of the socio-technical approach to irrigation research, which can show how technology acts as a controlling and mediating factor between biophysical and societal conditions. It is also useful as a method to identify stakeholders, their work practices and their interactions. The socio-technical approach to irrigation research is directed at understanding how irrigation systems are designed, operated and used by people to provide water. It focuses on three areas of stakeholder actions: the social construction of technology, social conditions of use and social impacts. Other research concepts include the study of domains of interaction in irrigation management – strategic interfaces where people come together to determine water supply. These are emergent and often specific to individual irrigation systems. This broad approach has proved particularly helpful for an adaptive design consistent with the knowledge and preferences of farmers.

APPROACH

Action research in irrigation development or reform nearly always involves the researcher in a *participation complex*, where the researcher has to work in (1) different domains of participation and (2) different development contexts of participation. In the first, key interfaces with different stakeholders must be understood, in terms of their sphere of influence and local representation, their interests in participation and their practice in relation to water supply and water users. In a large irrigation system, participation is not only with farmers directly. More commonly, negotiation will also be done through water user organizations (whose representation often changes), as well as with system operators and the contractors who often implement new construction. These different domains present different opportunities and challenges to participatory approaches.

Successes in some areas, such as working with farmers to agree new designs and irrigation schedules, may be tempered by problems in other areas, such as failing to get good quality construction under contracts. The researcher (and designer or implementer) has to work across these domains and interfaces – to get farmers' ideas put into practical reality. The different development contexts of participation have different concepts of innovation and different sets of participatory methodologies linked with them, as outlined below. These different development contexts can all occur in one programme – although often one may dominate depending on the objectives of projects and individuals. Conscious recognition of points of consensus and differences on participation-thinking can also help programme design and build better action between stakeholders.

Development Context 1: economic development and modernization

Participation is an approach (by agencies) to induce increases in performance or impact through providing conditions or incentives that enable farmers to take on new responsibilities and opportunities. Innovation is new activities that improve linkages between resource use and production – new techniques, artefacts or institutional relations that increase productivity, efficiency and economic returns, or reduce wastage and degradation. The participatory methodologies, criteria and activities (often more 'criteria focused') are:

- RRA;
- problem inventory;
- village credit camps;
- beneficiary targeting;
- participatory irrigation management (PIM);
- cost-sharing;
- accountability mechanisms;
- user-manageable design;
- on-farm trials;
- capacity building;
- farmer-to-farmer training.

This is often a context of induced innovation and needs participatory approaches that allow local negotiation and evolution of change, rather than blueprint models.

Development Context 2: joint planning and problem-solving

Participation is a process through which stakeholders influence, share control and work together to achieve desired change. Innovation is shown through the changed behaviour of the people involved and the sharing of knowledge and skills.

The participatory methodologies, criteria, activities (often more 'methodology focused') are:

- PRA;
- collegiate engineering;
- 'process' project planning;
- managing knowledge systems;
- consensus building and knowledge sharing;
- demand-driven development intervention;
- participatory technology development;
- networking to build platforms;
- stakeholder identification and interaction mapping.

This context works best at a small scale, and may only work where consensus is possible.

Development Context 3: social inclusion, improved equity and reduced vulnerability

Participation is organized effort to increase control over resources and regulative institutions in given situations on the part of groups and movements of those hitherto excluded (a definition from an Swiss International Labour Office programme).

Innovation is the delivery of different benefits to different people. The participatory methodologies, criteria and actions (often more 'action-focused') include the methods of point (2) above, but also 'empowerment principles' as summarized below:

- conscientization of farmers and representatives;
- a deliberately long time frame;
- working with local 'advisors' that have continuity, as well as current stakeholders;
- capacity building and user-focused design highlighting equity and basic water needs;
- uses a range of kinds of contact (not just PRA group exercises or consensus workshops);
- political and legal action for excluded groups;
- conflict removal as well as consensus building (recognize that consensus may be impossible);
- tolerate/recognize pluralism in many areas – law, science, technology use;
- work with local practices to adapt known science and technology;
- keep construction controlled or strongly supervised by the users.

This context recognizes the tensions and complex politics of negotiating change in many different arenas but needs highly motivated and conscientized actors to empower change.

Reflections

Development intervention is said to have three practical needs–explanation of the development problematic, information on which to develop action and conceptual tools for designing action. We need to think about typologies of action and not just methodologies for designing action to bring farmers' requirements into reality.

The socio-technical approach – and the concept of a 'participation complex' in action research – helps the design of participatory research and enables researchers to think about both the 'problem environment' and the 'project environment' in which they are working.

These shape the participatory methods that people can use and the emergent challenges of working with different stakeholders. Conscious recognition of different methods to achieve objectives – and the problems that may come from failures in any area and reasons for them – can help to actualize desired and agreed innovations.

LANDSCAPE LEVEL IN ECOSYSTEMS

Uncontrollable seasonal flooding affects over 10 million hectares, or 15 per cent, of the total rice land in south and south east Asia. During the dry season, land ownership is fixed according to tenure arrangements. In the rainy season, farmers grow deepwater rice and capture fish in the flood-prone areas. At this time, fish are considered a common resource, and community members are traditionally allowed access to private property for fishing.

Since 1997, ICLARM and the International Rice Research Institute (IRRI) have been undertaking an interdisciplinary and PAR project. The project is being implemented in collaboration with various governmental organizations and NGOs. The aim is to increase and sustain the productivity of rice and fish in the seasonally flooded ecosystem in Bangladesh and Vietnam as a demonstration for the entire region. The project strategy combines indigenous resource management techniques with semi-intensive fish culture and management technologies for the increased income of normal households.

APPROACH

The unit of analysis used is the resource management domain (RMD) at the landscape level. The RMD covers the environmental, social and economic characteristics of a recognizable unit of land and takes into account its inherent natural variability. The steps followed in the participatory problem analysis are to:

- collect and analyse secondary data;
- conduct a diagnostic field survey;
- conduct baseline socioeconomic, biophysical and institutional surveys;

- analyse data;
- present data;
- hold group discussions involving users, researchers and NGO representatives;
- identify problems and potential solutions.

The users/stockholders include landowners and other community members, at all levels, reliant on the landscape for fishing during the rainy season. To identify its clients, the project held meetings with farmers from different wealth groups, landless labourers and members of local organizations. The steps followed in assessing users' needs were (a) scientists and representatives from local-level organizations conducted a diagnostic survey, (b) baseline surveys were made of socioeconomic, institutional and biophysical conditions, and (c) group discussions were held with users. A main objective of the baseline survey is to later enable analyses of the project impact over time.

The concept of managed fish culture in deepwater rice fields is new. Thus, researchers designed technical options consulting users on their needs and taking into consideration their indigenous knowledge. Small-scale experiments were first initiated in Vietnam to show the potential of the technical options. These initial trials were then used to generate discussions between researchers and users about various aspects of trials to fine-tune the technical options. Users tested site-specific technical options with minimum support from researchers during 1997–2000. The project provided financing support, as seed money, during the first two years to cover material costs. Users deposited part of the proceeds from the experiments (eg, fish sales) to cover future project expenditures.

Researchers monitored water and soil quality, profitability, input use, fish consumption, group performance and sharing arrangements. Based on this information, the project analysed the impact of the technological innovations and the project processes. The results indicate that community-based fish culture in flood-prone ecosystems in Bangladesh and Vietnam is technically feasible, economically profitable, environmentally non-destructive, and socially acceptable. For the overall system, an additional income of US$150 per hectare in southern Vietnam to US$690 per hectare per year in Bangladesh is achieved, which is an increase of 20 to 160 per cent over the previous profitability.

Reflections

Deficiencies of the process are that it is researcher-initiatied and -dominated and not very participatory in areas where users have only a limited knowledge of the subject. On the other hand, in subsequent years the technology has been copied by neighbouring communities but often with differing arrangements. The approach does not work well in areas where group action is not viewed positively. The experiments are on an appropriate scale for representing the real world situation, and thus may be used for up scaling. The design and testing of the technological options included user participation. As regards sustainability, the community is less dependent on the project for funds and has an arrangement for group saving.

Problem analysis using the landscape-level resource management domain (RMD) as a unit has provided a better understanding of the integration of the biophysical and socioeconomic factors. A project implementation committee was established at each project site, including representatives from each user group. The committee oversees project implementation, prepares budgets, manages project accounts, negotiates sharing agreements (including participating members, responsibilities, access to the wild fish in the flood period, necessary guarding duties, etc), settles conflicts, supervises fish sales and distributes the proceeds from experiments. With support from researchers and NGO staff, different user groups have designed their own organizational arrangements for community-based fish culture in flood-prone rice ecosystems.

WATER MANAGEMENT IN AGRICULTURAL DEVELOPMENT

Different modes of NRM have far-reaching consequences for the well-being of poor farmers. The variation is clear indeed for irrigation, as illustrated for South Africa. In the arid and semi-arid regions of South Africa, irrigation is key to increasing the agricultural productivity and incomes of farmers. White farmers on large-scale private holdings use most of the country's irrigation water. The former Homelands, where most black South African farmers live, have only limited access to irrigation. A few hundred irrigation schemes were developed under the apartheid policy. However, parastatal agencies and private companies derived income from these schemes and dominated agricultural operations and water management. Poor black farmers received few benefits and were excluded from decision-making.

Since 1994, the new government has reversed apartheid policy and withdrawn agricultural support for schemes such as Arabie-Olifants. Black farmers, who are mostly women, are now expected to 'stand on their own feet'. This sudden change has led most households to abandon irrigated agriculture altogether, with negative impacts on their income and well-being. Few households were able to find alternatives to previous sources of credit, ploughing services and access to markets. Those few who returned to agriculture were often frustrated by breakdowns in the irrigation infrastructure.

The government has also started privatizing the ownership of the irrigation infrastructure and all rights and financial and managerial responsibilities for water management. Current users, or anyone interested, can buy. However, smallholders in these schemes, who already lack access to other inputs and markets, have no capital for buying. Instead, a small entrepreneurial 'elite', whose members are relatively well off, literate, mobile, male and well connected to policy networks, tend to be the owners of the new irrigation infrastructure. In such cases, the terms of production for the former producers become even more disadvantageous.

APPROACH

Worldwide, IWMI has conducted comparative research on participatory irrigation management in state-supported schemes. International and South African research findings have been input in a continuous dialogue with government officials and other actors influencing policy in South Africa. International comparison showed that the above-described pattern is typical in government-supported smallholder schemes in developing countries that are brusquely abandoned. The first generic conclusion is that irrigation management transfer in formerly state-supported schemes can only work if smallholder irrigated agriculture is profitable enough to bear the extra burdens of scheme operation and maintenance. So forging forward and backward linkages, credits and training are essential components, besides transfer of management, of an overall policy for the uplift of smallholder irrigation. Second, the process of transfer should be gradual; while inclusive, accountable member organizations are to be created.

Policy workshops and international comparison have led the government to launch an encompassing national policy for water for rural poverty alleviation and the re-vitalization of smallholder irrigation schemes. The IWMI's research serves as an intermediate stage between the key stakeholders in state-supported irrigation, linking poor farmers and irrigation policy-makers and implementing agencies.

Reflections

Comparative research on gender in irrigation in South Africa identified issues that will be relevant for the future creation of inclusive water users' associations. First, most farm decision-makers in smallholder schemes are women; irrigated agriculture is a female farming system. This is related to the cultural division of tasks within households and the state's gendered off-farm employment and homeland policies. Second, although some women have plots in their own names, others have long-life tenure security to the land of their male kin. Elsewhere in irrigation management, water rights and membership of water users' associations are commonly vested in landowners, rather than land users. However, the National Water Act (1998) of South Africa leaves the option open to disconnect water rights from primary land titles. Hence, farm producers and users of a portion of land can become members of the new water users' associations. This would especially benefit women farmers. From both a gender and poverty perspective, the challenge is to organize producers, irrespective of type of land rights, in a bottom-up way, and to ensure transparent elections of committees that remain accountable to their constituencies.

SOIL AND WATER CONSERVATION IN RURAL PRODUCTION

The Promotion of Soil Conservation and Rural Production (PROSCARP) is a national-level project aimed at improving the nutrition and health of smallholder farmers in Malawi. Since 1995 it has worked through Catchment Area Development Committees (CADCs), which have mostly been elected in village meetings and which cover a defined area. In 1999 these numbered over 300, scattered throughout the country.

The project aims to work in a participatory mode with the CADCs and with other agencies. This is recognizably difficult because of the large size of the project and because of the historical legacy of a national extension system that has been top-down and, while it has adopted a participatory ethos, it is currently underresourced and demoralized. The project offers a range of soil- and water-related technology options to farmers, including realignment of ridges on the contour, vetiver grass, green manure crops, legume rotation crops, agroforestry and minimum tillage. In order to encourage uptake, the project has also used incentives such as providing free seeds and seeds on loan, paying for labour on vetiver grass nurseries and providing village wells and pit latrines.

Approach

During 1999, a small study team met with the CADC members, local extension staff and other members of the community over periods of 1–3 days per area. Discussions were held with individuals and with focus groups and included some ranking and time-line exercises. Focus groups (village headmen, male and female CADC members and other villagers) were encouraged to share their findings with each other and to discuss further the implications for the soil and water conservation programme in their area.

The use of various soil and water conservation technologies has evolved over time. Village headmen who remembered events from the 1940s were able to describe and quantify some of the main changes in physical soil conservation and this exercise helped to place the current activities of the CADCs within a historical context. A notable feature of the changing soil and water conservation practices was a focus on physical structures at field level initiated by external agencies. This is not surprising, given that during the colonial period the village headmen were instructed to oversee and enforce the construction of soil and water structures. However, it raised questions about how conservation technologies were introduced and perceived. Were the changes largely a response to external threats and incentives? Were physical structures regarded more as significant innovations than changing cultural practices? A discussion around why some people had realigned ridges, while others had not, provided insights into how the local CADC operated and the role of incentives in the uptake of new conservation technologies.

Three main reasons accounted for the slow spread of technologies. First, a technology was adopted in order to receive an incentive. Second, difficulties occurred in access to technologies and the licence to distribute those that were available.

The technologies such as vetiver grass were seen to be the property of the project and its CADCs, and therefore it was felt that approval from above might be needed in order to pass these on to communities outside of the defined catchment area. Third, some of the technologies had not yet shown clear benefits and others were not suited to local conditions.

The CADCs were found to function rather differently in each area, depending on how they had been established and how the local people (including local extension agents) had taken them forward and interpreted instructions from the project. Some had been negatively affected by village politics, a fact of life in rural Malawi. The CADCs were most functional when they had effectively incorporated local village leadership, and least so when used as a vehicle for one faction to challenge the current village leadership. This further raised the issue that committees based on support from local political positions (based on people groups) often did not correspond with ideal geographical areas for integrated (above field level) soil and water management.

Most local extension workers acknowledge that they spent a disproportionate amount of their time on project-related work, rather than on their general duties. They justified this in terms of project resources and farmer interest in the CADC areas. They also noted that they had impossibly large areas to cover and even without a more intensive project would not be able to cover all their mandate area. This raised the issue of the potential side effects of introducing more intensive extension approaches in the context of an expanding rural population and a shrinking number of extension staff, lacking in mobility.

The donors, concerned that insufficient project resources were going directly to the community, raised the issue of how to stimulate soil and water conservation in a sustainable manner. More direct payments to communities may have the effect of being a disincentive to neighbouring communities not directly involved but potentially benefiting through extension and farmer-to-farmer dissemination. It was recognized that extension staff needed to be rewarded for their good work as facilitators and encouragers. The question is raised as to the compatibility of the two project objectives of poverty alleviation (in the short term) on the one hand and soil and water conservation that is both participatory and sustainable (in the longer term) on the other.

Reflections

Several issues of concern have arisen during implementation of the project including:

- The spread of technologies to neighbouring communities has been much slower than expected, limiting geographical impact.
- Information on how the individual CADCs function on a day-to-day basis is not comprehensive but there are indications that some CADCs are perceived as operating like exclusive clubs – restricting benefits to a few community members.
- The local extension workers may be spending too much (from the perspective of their line managers) or too little (from the project perspective) time in the catchment areas.
- A high proportion of project costs is spent on administration and related recurrent costs (travel and field allowances) compared to the amount that reaches the communities directly.
- Provision of incentives to targeted groups has discouraged technology uptake by a wider group of farmers over a larger geographical area.

The project management is concerned that implementation should be more participatory. Thus studies at village level using PRA tools were conducted to explore this point and the above concerns. Expectations were raised, both for villagers and local field staff, during meetings. It is unclear whether the project will have the capacity to address the issues raised, particularly the plans of the local community and field staff to further expand the project's geographical scope in response to demand from neighbouring communities who have perceived the benefits. Despite these concerns, benefits accrued from the project. Many farmers were exposed to a range of technologies, and a significant number of them received benefits in terms of cash payments and increased production from using the technologies provided. Front line staff increased their technical skills and skills in working through local committees and training farmers to train other farmers. Farmers were trained in soil and water conservation techniques. The village-level studies placed project activities in a historical context, both for the local community and for the project staff.

RESOURCE ISSUES IN INTEGRATED WATERSHED MANAGEMENT

This case study attempts to integrate sustainable principles into watershed management in two Nepalese watersheds (Jhikhu Khola and Yarsha Khola). These Middle Mountain watersheds are some of the most intensively used landscapes on earth and exhibit all the resource problems that are now of major concern in developing countries. Water shortages, water pollution, soil fertility deterioration, deforestation, lack of animal feed, stagnating biomass production, inequity, poor food security, poverty, increasing workload for women, few alternative economic options and poor infrastructure support are all part of the overall problem being addressed.

The challenge is to arrive at methods that are adaptable to complex conditions and that facilitate integration and interdisciplinary activities, and build linkages between researchers, farmers and local and national institutions. Access to long-term funding from IDRC (and more recently SDC) over an eight-year period has allowed us to develop a comprehensive resource database for the watershed and helped make the transition from a basic, science-driven project to one that is primarily participatory. We use the watershed (not the community) as our unifying unit for research because we can model landscapes, water, sediment, nutrient dynamics and climatic change effects at that scale.

Approach

The community-based NRM approach (CBNRM) begins with a rapid PRA to identify common concerns and issues in the communities within the watershed. Then a GIS database is built consisting of geology, soils, topography and land use layers. In the field we establish monitoring stations for climate, hydrometry, soil fertility and soil erosion. The issues raised by the communities are addressed using a GIS approach that includes overlay stratification, modelling, statistics and socioeconomic surveys. Key factors indicative of climatic conditions (elevation and aspect), the dominant bedrock and superficial material types and dominant land uses are identified. These are then divided into unique categories (two for elevation, two for aspect, three or four for contrasting rock type and usually four for land use type). The combination of these factors plays the dominant role in shaping and using the landscape and this 2×2×3×4-combination matrix is then used to divide the landscape into 48 possible landscape combinations. The GIS overlay technique is applied to show the dimension and location of each combination. Ten farmers and ten members of forest user groups are then randomly selected in each of these 48 classes of landscape and a participatory survey is conducted. Samples of the dominant soils in each chosen farm or forest are collected and analysed and information is obtained on farm and soil inputs, production, socioeconomics, gender and equity and forest use and management practices.

Collaborative participatory farm interventions are then initiated to address the farmers' concerns. Based on this approach we have identified that only one-third of all farmers apply enough nutrients (N and P) to a maize crop in double rotation and in these farming systems the long-term soil fertility is not sustained. At the same time, we can show that the nutrient deficits in an irrigated crop rotation system of rice, wheat and maize is only prevalent in about 40 per cent of all farms. This can now be expanded to other cropping rotations by examining the nutrient balance situation and economic consideration when cash crops such as potatoes and tomatoes are introduced into the rotation. We can also apply scenarios to these systems and simulate possible outcomes.

Our most interesting and most challenging research lies in how to correct problems and how to rehabilitate sites, not just in identifying and quantifying the problem, or in determining the rate of degradation. Highly degraded sites often occur on common land where the prospects for rehabilitation are poor because of the great effort needed to establish biomass and the low possibility of short-term economic returns. However, we demonstrated that up to 40 per cent of the total annual sediment load in the river originates from such sites and the impacts on irrigation systems downstream are large. These areas provide an opportunity for researching how to develop community forests and grazing lands that eventually can become biodiversity gardens.

Reflections

The methodologies used are more complex and require newer skills than are traditionally available at educational institutions. Thus, much effort has to be spent on training and education. This delays diagnostic and intervention research but has more long-term benefits. Working in an interdisciplinary manner is also much more difficult and demanding. Putting together the right team configuration and matching it with the right personalities is probably the biggest challenge. The approach is highly time consuming, thus project funding should be assured for longer time periods than the traditional 3–5 years. The focus tends to be around communities at the expense of integration within larger more natural units such as watersheds and coastal or ecological zones. Another disadvantage is that scaling up cannot easily be accomplished when not incorporated into the research activities at the outset. The advantage of this method is that it covers all environmental conditions in the watershed and enables us to determine how much each individual factor contributes to the overall variance of key resource indicators. Based on this sampling design, the analysis of variance or non-parametric significance tests can be used in quantifying potential causes and rates of degradation. The social factors are not stratified in a statistical manner but, because we look at all types of biophysical conditions, we capture members from most of the socioeconomic spectrum in the watershed. The initial effort is large but the payback is enormous because this type of survey can be done once every 5–7 years in the same watershed. Provided all the information is geo-referenced, it gives us the opportunity to document the dynamics of land use, soil fertility and socioeconomic conditions.

The CBNRM approach addresses the immediate concerns and issues of the community, providing a better forum for communication among researchers, community participants and the general public. The focus is on the poorer fraction of the society and this allows more emphasis to be placed on issues of gender and ethnicity. Gender is of particular interest because the workload of women in Nepal has increased with agricultural intensification and off-farm employment by men. Conducting research that focuses on reducing the workload of women (eg, improving fodder and fuelwood supplies, facilitating access to safe drinking water, improving soil nutrient management) can therefore have immediate effects in improving the livelihood of rural families. Well-chosen interventions and dissemination of successful results can be facilitated and applied more rapidly using a CBNRM approach. Issues are addressed in a more interdisciplinary manner and this should lead to a better understanding of the environmental system and result in more holistic and permanent solutions. The involvement of stakeholders that play an active part in the research provides a reality check on the relevance of the research. The approach can facilitate conflict resolution because stakeholders are incorporated into the research from the start. It leads to more effective public education and forces researchers to communicate better with the public in explaining why the research is important and what the results mean. The act of supporting credible research helps build intellectual and scientific legitimacy for political reform. The CBNRM approach has stimulated internal discussions leading to more open policies. It is flexible and can readily be adjusted to a wide range of conditions.

THE EVOLUTION OF COMMUNITY FORESTRY IN PRACTICE

Early initiatives understandably tended to become focused on those issues perceived to be of particular importance. Of these, the fuelwood shortage became far the most important. It has been estimated that wood provides roughly 20% of all energy in Asia and Latin America, and that about 50% of all energy in Africa is wood generated. The early analyses emphasised the huge numbers of people affected, the apparent "gap" between demand for fuelwood and sustainable supplies - and a seemingly exponential growth in demand with population growth - and the growing burden placed on users of having to search ever further afield for fuelwood and of having to divert crop and animal residues needed for soil working or as livestock feed to fuel use.

A general conclusion from this earlier analytical work was that existing wood stocks were being widely mined to meet fuelwood demand, that there was no feasible large scale alternative to wood fuel - other than other biomass such as crop residues and dung - and that the principal means of averting growing shortages, and the attendant deforestation and human suffering, was to initiate widespread planting of additional trees. When applied to individual countries and regions such analyses resulted in programme targets on a formidable scale. For example, a major World Bank study for sub-Saharan Africa estimated that tree planting would have to increase fifteen-fold in order to close the projected fuelwood gap in the year 2000. As a consequence, a very large part of the initial investment in community forestry was in the form of afforestation projects to increase fuelwood supplies. Many early community forestry projects therefore took shape more as a response to an energy supply/ demand problem, rather than to the problem of meeting local needs for trees and tree products.

The design of early community forestry projects was also strongly influenced by the fact that the more striking existing programmes were organised in the form of communal activities - the village woodlots in Korea, the panchayat woodlots of early social forestry programmes in India, the village afforestation programme in Tanzania. As is suggested by their titles, all of these were concerned with creating new plantations, rather than with management of existing forests.

These perceived imperatives had the effect of concentrating the early community forestry effort to just a rather narrow part of the spectrum of linkages between people and trees and tree products that had been identified initially - namely to establishment of new plantations and to fuelwood. There was little in the first generation of projects which was concerned with outputs from existing forests, or with the food, employment and income dimensions. Indeed, the concern with meeting subsistence needs for fuelwood even led, on occasion, to attempts to exclude income generation activities from the project design on the grounds that it was inconsistent with the perceived subsistence aims of community forestry.

BROADENING THE KNOWLEDGE BASE

In subsequent sections of this document we explore the reasons why these, and other, unforeseen developments occurred in the early years of community forestry. However, a number of salient lessons which began to emerge early in the period deserve to be mentioned at this point.

One is that production and use of tree products at the village level is in practice often embedded in complex resource and social systems, within which most of the factors that affect our ability to intervene with forestry solutions are of a non-forestry nature. They are primarily human factors, connected with the ways women and men organize the use of their land and other resources. They therefore require situation-specific approaches and are unlikely to be successfully tackled by generalised solutions or approaches that address only a single element of the situation.

A second is that earlier analyses of the nature of women's and men's dependence on trees and tree products was in some respects incorrect or incomplete and the solutions identified were consequently inappropriate. As is discussed in some detail later, this is particularly the case with respect to responses to declining supplies of fuelwood and to the attempts to insert interventions that conflicted with the existing social and institutional framework within a community. A third is that even projects which have sought to identify local needs, aspirations and possibilities have in practice done so more on the basis of the views of planners and others from outside than on the local people themselves. Dialogue to achieve local participation has all too often started only after the project design has been finalised and is in place. Though the concept of participation took root quickly, in practice it has been, and still is, more frequently preached than practised.

A fourth is that "community forestry" has suffered from considerable confusion and lack of clarity as to its nature and purpose. The use of this umbrella term seems on occasion to have obscured the fact that the objectives set for projects to support community forestry have varied considerably. Project design, and performance, have frequently suffered from a lack of clarity as to which of these objectives were being pursued or had priority. Although some among multiple goals may be congruent or reinforce each other, others may be in conflict. Planting trees to meet environmental objectives such as soil protection is unlikely to produce sufficient output of saleable products as to be economically attractive to the farmer. Similarly, tree growing designed to generate income is unlikely in itself to benefit those with little or no land. Production to meet both subsistence and market needs is unlikely to be achieved with a single production model. Projects originally designed to meet a production goal are unlikely to be equally successful at achieving a subsequently added social goal, such as favouring the poor, unless they are appropriately restructured. And so on.

Index